computer applications in the earth sciences

MATHEMATICAL MODELS OF SEDIMENTARY PROCESSES

an international symposium

COMPUTER APPLICATIONS IN THE EARTH SCIENCES
A series edited by Daniel F. Merriam

1969 – Computer Applications in the Earth Sciences
1970 – Geostatistics
1972 – Mathematical Models of Sedimentary Processes

MATHEMATICAL MODELS OF SEDIMENTARY PROCESSES

an international symposium

Proceedings of an international symposium held at the VIII International Sedimentological Congress in Heidelberg, West Germany, on 31 August 1971. The meeting was cosponsored by the International Association for Mathematical Geology.

Edited by Daniel F. Merriam
Jessie Page Heroy Professor of Geology and Chairman
Department of Geology, Syracuse University
Syracuse, New York

PLENUM PRESS · NEW YORK – LONDON · 1972

Library of Congress Catalog Card Number 72-78629
ISBN 0-306-30701-4

A Division of Plenum Publishing Corporation
227 West 17th Street, New York, N.Y. 10011

United Kingdom edition published by Plenum Press, London
A Division of Plenum Publishing Company, Ltd.
Davis House (4th Floor), 8 Scrubs Lane, Harlesden, London, NW10 6SE, England

Printed in the United States of America

To all future sedimentologists
may they benefit as a result of this pioneering effort

PREFACE

This volume reports the results of a symposium held in Heidelberg during the International Sedimentological Congress in late August and early September, 1971. The symposium, co-sponsored by the International Association for Mathematical Geology, entertained the subject, "Mathematical Models of Sedimentary Processes."

The subject is most appropriate because sedimentologists have long been concerned with processes and mechanisms of sediment dispersal. Much effort has gone into building physical models such as flumes, stream tables, wave tanks, wind tunnels, etc., to help understand sedimentological processes. Quantitative methods (especially statistics) have been utilized in summarizing these data. It is timely then with the recent developments of simulation and application of computer techniques that a symposium be addressed to the use of "Mathematical Models of Sedimentary Processes" involving some of these new statistically oriented methods and available data bases.

Experimentation in geology has been hampered by a scale factor. That is, it is difficult to find suitable materials for physical models; it is difficult to find a mechanical device which properly represents the forces involved; it is almost impossible to allow adequately for geologic time. Statistically valid models are difficult to obtain with physical models because of material replicate problems. Most problems including the time factor, however, can be eliminated with mathematical models.

Mathematical models can be infinitely varied in any number of combinations easily and quickly with the computer. After development of a computer program, cost of experimentation is small. The stochastic element is easily taken into consideration. Repeated experiments can be accomplished with accuracy by different investigators. In summary, mathematical modeling offers many advantages in research, and simulation promises some answers to previously unsolved problems.

With the advantages and possibilities in this new area of study, it was thought prudent to bring together experts to exchange information and discuss problems. A symposium was judged the appropriate method in which to accomplish this objective and the IAS Congress provided the vehicle for interdisciplinary and international cooperation. Sedimentologists, hydrologists, geographers, and statisticians from eight countries contributed 13 papers to this volume. At the meeting only nine of the announced papers were presented orally. Unfortunately, one of those presented was not available to be included in the proceedings. Two other papers presented at other sessions, however, are included.

The background and fundamentals for mathematical models in the earth sciences is outlined in a review paper by G.F. Bonham-Carter. He invokes concepts and ideas from a recent book he coauthored with J.W. Harbaugh (Computer Simulation in Geology, John Wiley & Sons, New York, 1970), and thereby sets the stage for those interested but not yet involved. The papers can be conveniently grouped into categories depending on interests. Several are concerned with vertical successions, their origin and correlation; others deal with areal distributions and variation. The papers give some idea as to the status of quantification attained in geological research. Many techniques would be impossible to apply without the computer. Although geomathematics and computer applications are relatively new, their future seems assured and results should be rewarding.

Many people helped make the symposium successful. First of all, the participants are to be thanked for their contribution. Secondly, arrangements in Heidelberg for the meeting were made by Professor German Muller of the Mineralogy-Petrography Institute, University of Heidelberg. Thirdly, the manuscripts were typed and proofread by Miss Grace Hillman, Department of Geology at Syracuse University. And last, the cooperation and help of Plenum Publishing Corp., Robert N. Ubell, Editor-in-Chief, is to be acknowledged.

It is hoped that this publication will adequately represent the air of interest and enthusiasm as shown by the participants in Heidelberg. Unfortunately, the discussion could not be reproduced here but the reader is encouraged to pursue the subject further either through contact directly with the authors or in the literature (a large volume on the subject is now being generated).

Again, a thanks to all those who took part in the symposium and best wishes to those taking part in absentia by reading the proceedings.

Heidelberg, West Germany
September 1971

D.F. Merriam

LIST OF PARTICIPANTS

G.F. Bonham-Carter, Department of Geology, University of Rochester, Rochester, New York, USA

J.M. Cocke, Department of Geology, East Tennessee State University, Johnson City, Tennessee, USA

J.C. Davis, Kansas Geological Survey, The University of Kansas, Lawrence, Kansas, USA

G. de Marsily, Laboratoire d'Hydrogeologie Mathematique, Ecole Nationale Superieure des Mines de Paris, Fontainebleau, France

F. Demirmen, N.V. Turkse Shell, Ankara, Turkey

C. Dumitriu, Geological Institute, Bucharest, Romania

M. Dumitriu, Geological Institute, Bucharest, Romania

N.E. Hardy, Department of Geography, The University of Kansas, Lawrence, Kansas, USA

K.I. Heiskanen, Geological Institute, Karelian Branch of Sciences, Petrosavodsk, USSR

J. Jacod, Centre de Morphologie Mathematique, Ecole Nationale Superieure des Mines, Fontainebleau, France

P. Joathon, Centre de Morphologie Mathematique, Ecole Nationale Superieure des Mines, Fontainebleau, France

W.C. Krumbein, Department of Geological Sciences, Northwestern University, Evanston, Illinois, USA

W.O. Lockman, Department of Geography, The University of Kansas, Lawrence, Kansas, USA

M.J. McCullagh, Kansas Geological Survey, The University of Kansas, Lawrence, Kansas, USA

D. Marsal, Gewerkschaften Brigitta und Elwerath Erdgas Erdol, Hannover, FRD

D.F. Merriam, Department of Geology, Syracuse University, Syracuse, New York, USA

W.A. Read, Institute of Geological Sciences, Edinburgh, Scotland, UK, and Kansas Geological Survey, The University of Kansas, Lawrence, Kansas, USA

R.A. Reyment, Paleontologiska Institutet, Uppsala Universitet, Uppsala, Sweden

W. Schwarzacher, Department of Geology, The Queen's University, Belfast, Northern Ireland, UK

CONTENTS

Optimization criteria for mathematical models used in sedimentology, by G.F. Bonham-Carter.................... 1

Interpretation of complex lithologic successions by substitutability analysis, by J.C. Davis and J.M. Cocke.................................... 27

Mathematical models for hydrologic processes, by G. de Marsily.................................. 53

Mathematical search procedures in facies modeling in sedimentary rocks, by F. Demirmen...................... 81

Monte Carlo simulation of some flysch deposits from the East Carpathians, by M. Dumitriu and C. Dumitriu........ 115

Diffusion model of sedimentation from turbulent flow, by K.I. Heiskanen................................. 125

Conditional simulation of sedimentary cycles in three dimensions, by J. Jacod and P. Joathon................. 139

Areal variation and statistical correlation, by W.C. Krumbein.................................. 167

Formation and migration of sand dunes: a simulation of their effect in the sedimentary environment, by M.J. McCullagh, N.E. Hardy, and W.O. Lockman......... 175

Mathematical models for solution rates of different-sized particles in liquids, by D. Marsal............... 191

A simple quantitative technique for comparing cyclically deposited successions, by W.A. Read and D.F. Merriam.... 203

Models for studying the occurrence of lead and zinc in a deltaic environment, by R.A. Reyment................. 233

The semi-Markov process as a general sedimentation model, by W. Schwarzacher................................ 247

Index... 269

OPTIMIZATION CRITERIA FOR MATHEMATICAL MODELS USED IN SEDIMENTOLOGY

Graeme Bonham-Carter

University of Rochester

ABSTRACT

Optimization principles play a key role in mathematical sedimentology in at least three ways.

(1) The data-gathering process is optimized so as either to maximize the amount of information to be obtained with limited cost of sampling and measurement or to minimize the cost of obtaining at least a certain known amount of information.

(2) Many commonly used data-analysis models have been formulated using optimization criteria. Examples include: regression by minimizing the sum of squares; principal component analysis by maximizing the amount of variation associated with each successive component axis, subject to constraints; discriminant-function analysis by placing decision surfaces so as to maximize the separation of predefined groups; cluster analysis by maximizing the compactness of groups; and several others.

(3) Simulation modeling may involve optimization. First, principles of least work or similar optimization criteria may provide keys to the mathematical formulation of a model. This follows as a consequence of the optimization inherent in nature. Second, a basic goal in simulation is to adjust the parameters so as to minimize the difference between the model's output and the real-world response. Furthermore, the process of exploring the sensitivity of a simulation model to systematic changes in parameters is a natural extension of optimization.

An appreciation of basic optimization principles leads to a clearer understanding as to how mathematics can help to solve sedimentological problems.

INTRODUCTION

The concept of optimization is recognized widely in natural science, engineering, business administration, and many other fields. In each situation, optimality principles are important components in the design of systems. In physical, chemical, and biological phenomena, the design of natural systems may be optimal with respect to a cost function. Similarly, the artificial systems constructed by engineers and operations research analysts are designed to optimize their efficiency in terms of time or money.

Many important physical laws incorporate optimization criteria. Wilde and Beightler (1967) point out that the laws of statics were deduced from the principle of least restraint, stated by Gauss in 1829; and that in 1834 Hamilton formulated his famous minimization principle, from which could be obtained all the optical and mechanical laws then known, and which forms one of the fundamental elements of wave mechanics and relativity. Indeed, Goldstein (1950) points out that

> ...in almost every field of physics [optimization] principles can be used to express the "equations of motion"...Consequently, when [such a] principle is used as the basis of formulation, all fields will exhibit, at least to some degree, a structural analogy.

In chemistry, Le Chatelier's principle involves optimality. This principle can be stated thus: If a system is in equilibrium, a change in any of the factors determining the conditions of equilibrium will cause the equilibrium to shift in such a manner as to minimize the effect of this change. This principle is, of course, widely used in thermodynamics for studying the direction of chemical reactions.

The ubiquity of optimal designs in living organisms has been recognized by many authors (e.g. Rosen, 1967). It is no accident that bees build honeycomb cells with regular geometry, and that the branching of arteries and veins in mammals follows a definite pattern. The process of evolution by natural selection has been a natural optimization process, with survival only of the fittest.

Engineering, business administration, and other fields have common ground in the discipline known as operations research. The keystone of this subject is optimal design. Factories are designed to optimize production; military strategy is designed to maximize captured territory or minimize casulaties; supermarkets are designed to minimize queuing. In operations research, the objective function being optimized is usually simple, either money (cost or profit) or time. Nature's systems may not have obvious objectives, and the objective function may be difficult to discern.

In geology, we make use of physical, chemical, and biological principles, and their attendant optimal features. We may use optimal designs for sampling and data analysis. Furthermore, concepts akin to Le Chatelier's principle have been used to account for some geomorphological phenomena. For example, Langbein (1966) has shown how the geometry of rivers continually adjusts to changes in discharge in such a manner that variation of the parameters width, depth, and velocity is the least possible. Leopold and Langbein (1966) showed that meanders seem to be the form in which a river does the least work in turning. Optimality has been implictly recognized in sedimentology too. For example, in discussing current ripples, Bagnold (1956) said

> ...ripple marks represent a surface of greater friction, which is constructed spontaneously from cohesionless sand grains when shearing stress is applied more rapidly from the current to the bottom...the ripple creates bottom roughness and hence greater resistance; by this means they restore equilibrium.

The process of restoring equilibrium implies minimizing deviations from equilibrium and therefore optimization.

The purpose of this paper is to point out how optimization principles play an important role in mathematical modeling in sedimentology. The mathematical formulation of optimization is basically similar for differing applications. In other words, optimization is a useful general theme, because totally different phenomena may be analogous when viewed mathematically. The resulting insights can be most rewarding.

We will begin by examining the mathematical form of a well-known optimization problem; some applications of optimization of interest to sedimentology are then explored.

MATHEMATICAL FORM OF AN OPTIMIZATION PROBLEM

The problem is that of getting the most enjoyment from a party without going beyond the hangover limit. Figure 1, which is entirely hypothetical of course, is a graph whose axes correspond to the variables number of drinks (x_1), and alcohol content of

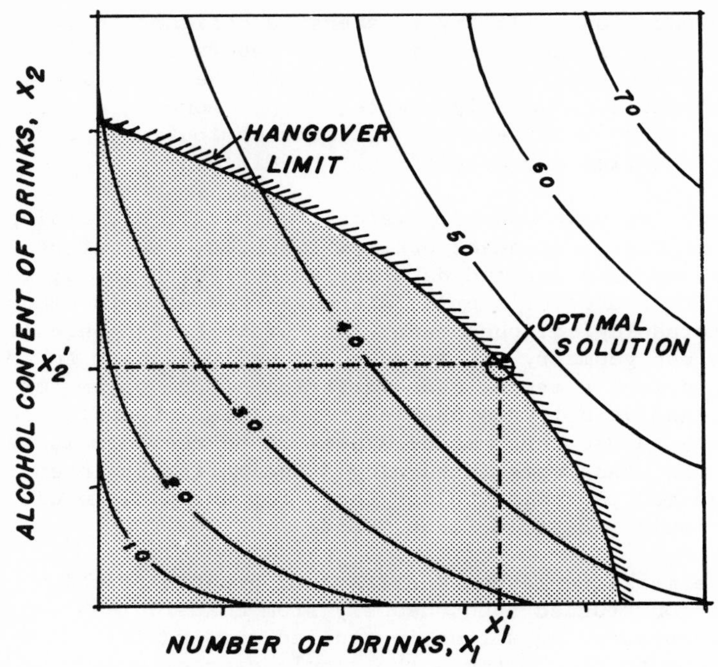

Figure 1. - Enjoyment contoured as function of decision variables, number of drinks, (x_1) and strength of drinks, (x_2). Shaded region is area of feasible solution, bounded by nonlinear constraint. Enjoyment is maximized at x_1^1 and x_2^1, optimal solution.

drinks (x_2). A third variable, termed enjoyment (z), which changes with both x_1 and x_2, is contoured as a surface on the graph, and is governed by the relation

$$z = f_1\ (x_1,\ x_2)\ .$$

This simply says that enjoyment is a function of the number and strength of the drinks! As the contours show, enjoyment increases with either an increase in the number of drinks, or an increase in alcohol content, or both.

There is a limit, however, as we all know too well. For some this may be the blood-alcohol level permitted for driving, for others it may be the hangover level. On the graph this is shown as a line which cuts across the contoured surface. The area above the line is beyond the limit; the area beneath the line is the area where enjoyment is within bounds. This constraint may be expressed as

$$c \geq f_2 (x_1, x_2)$$

which says that the hangover limit, c, must be greater than or equal to a function of the number and strength of drinks.

The problem is to decide which values of x_1 and x_2 will maximize enjoyment subject to the hangover constraint. This is a process of optimization. The number, x_1, and the strength, x_2, of drinks are the decision variables; enjoyment, z, is known as the objective function which is to be maximized (or in some situations minimized); the hangover constraint is an inequality which bounds the area of feasible solution (shaded on graph).

Any point in the area of feasible solution satisfies the hangover constraint; there is only one point in this area, however, at which enjoyment is maximized. This is the optimal solution, and the values of the decision variables for this point are x_1^1 and x_2^1, as shown on the graph.

In this example we have not specified the nature of the functions f_1 and f_2, nor have we defined the constant, c. These clearly would differ from one individual to the next, and would likely be complex.

The method of solution here is graphical, and this is possible if the number of decision variables is one or two. For three or more decision variables one must either turn to calculus methods, so-called hill-climbing methods, or a class of techniques known as mathematical programming (nothing to do with computer programming). For an introduction to some of these methods, see Harbaugh and Bonham-Carter (1970); for more advanced reading, see Wilde and Beightler (1967).

We will now illustrate the application of optimization principles to

(1) sampling strategies,
(2) data analysis, and
(3) computer simulation modeling.

OPTIMIZING SAMPLING PROCEDURES

Because of the variable nature of most sedimentological data (as compared with the relatively well-behaved variables of the chemist and physicist), the sedimentologist is faced with the necessity of collecting enormous amounts of information in order to obtain precise estimates for parameters of interest. In any particular study, the investigator must inevitably decide how many data points he needs. Usually this amounts to a trade-off of cost (time and money) versus the precision of the desired result.

The study of sampling techniques is an advanced field in statistics in its own right; there are several textbooks devoted exclusively to the subject (e.g. Cochran, 1963). Many authors have discussed sampling problems as related to sedimentology. Textbooks with chapters devoted to this topic include those by Krumbein and Graybill (1965) and Griffiths (1967); both of these books give comprehensive lists of references.

In terms of optimization theory, sampling procedures normally involve either (1) determining the number and distribution of samples that will minimize costs and yet achieve at least a basic minimum amount of information; or (2) determining how samples should be collected for a fixed cost so as to maximize the information gained.

Let us take a simple situation to illustrate this point. Suppose that the illite content of samples from an area of the continental shelf is under investigation. Preliminary studies show that the percentage of illite, x, is a normally distributed variable, and that the population variance, σ, equals 7.78 percent. An investigator wishes to determine the mean value, $\bar{x}$, of illite at a particular locality.

Because illite percent is a normally distributed variable, the population mean (μ) will lie within a known confidence interval, depending on the sample mean and the standard error of the mean, $\sigma_{\bar{x}}$. This can be stated as

$$P\left\{ \bar{x} - 1.96\sigma_{\bar{x}} < \mu < \bar{x} + 1.96\sigma_{\bar{x}} \right\} = .95.$$

This says that the probability of the true population mean lying between the indicated lower and upper limits is 95%. Or alternatively, the sample mean and the true mean differ by $\pm 1.96\sigma_{\bar{x}}$ with 95% probability. The number 1.96 is, of course, the value of the standard normal deviate, z, for a two-tailed hypothesis test at the 95% level of probability.

The standard error of the mean, $\sigma_{\bar{x}}$, depends closely on the number of samples. It has been shown that

$$\sigma_{\bar{x}} = \frac{\sigma}{\sqrt{n}} . \tag{1}$$

If we define d as half the 95% confidence interval for the mean, then

$$d = 1.96\sigma_{\bar{x}} ,$$

and substituting $\sigma_{\bar{x}}$ from (1) we obtain

$$d = 1.96 \frac{\sigma}{\sqrt{n}} ,$$

and we know that $\mu = \bar{x} \pm d$ with a probability of 95%.

Assume that each sample costs $3 to collect and analyze. Then

$$c = 3n ,$$

where c = total costs in dollars.

The investigator can now either (1) find the number of samples, n, that will minimize cost and yet give a known confidence range, d; or (2) find a number of samples that will minimize d yet cost no more than a fixed number of dollars.

For situation (1), suppose that $d \leq 3$, then

$$3 \geq \frac{1.96\sigma}{\sqrt{n}} .$$

Substituting $\sigma = 7.78$ and solving for n

$$n \geq \left[\frac{(1.96)(7.78)}{3}\right]^2 \geq 25.81 .$$

Because n must be a whole number, $n = 26$ is the value that minimizes cost (3 x 26 = $84) yet ensures that $\mu = \bar{x} \pm 3\%$ with 95% probability.

For situation (2), assume that no more than $90 can be spent on sampling. There is really no optimization involved here because if d is to be minimized it is clear that n must be as large as possible and therefore

$$n = \frac{90}{3} = 30 .$$

However, we can calculate the size of the confidence interval, d, corresponding to this sample size as

$$d = \frac{(1.96)(7.78)}{\sqrt{30}} = 2.78 .$$

Thus if no more than $90 are to be spent, 30 samples should be taken and this will permit determination of the mean illite content to within ±2.78% of the true mean at the 95% level of probability.

From this simple example, let us progress to an analysis reported by Kelley and McManus (1969) and Kelley (1971). These authors were designing an optimum sampling plan for studying the grain size at localities along the west coast of United States. At each sampling locality they could take one or more grab samples, and each grab sample could be split into one or more subsamples. They estimated the cost of each grab sample as $5 and the analysis of each subsample as $5. Thus their cost equation was

$$c_t = n_g c_g + n_g n_s c_s = 5n_g + 5n_g n_s \quad ,$$

where c_t = total cost of samples in dollars,

n_g = number of grab samples, and

n_s = number of subsamples per grab.

The problem was to estimate the average grain size, x (measured in phi units) at each station. How many grabs should be taken per station and how many subsamples should be analyzed from each grab?

In order to answer this question a detailed analysis of the variances due to grab sampling and subsampling was made. Then the standard error of the mean was expressed as

$$\sigma_{\bar{x}} = \frac{\sigma_g}{\sqrt{n_g}} + \frac{\sigma_s}{\sqrt{n_g n_s}} \quad ,$$

where $\sigma_{\bar{x}}$ = standard error of the mean of x,

σ_g = square root of population variance due to grab sampling

σ_s = square root of population variance due to subsampling.

Substituting estimated variances of $\sigma_g^2 = .0291$ and $\sigma_s^2 = .2275$ and squaring each term Kelley and McManus obtained the following expression

$$\sigma_{\bar{x}}^2 = \frac{.0291}{n_g} + \frac{.2275}{n_g n_s} \quad . \tag{2}$$

If d is again defined as half the confidence interval about the mean at 95% level of probability, $d = 1.96\sigma_{\bar{x}}$, then

$$\sigma_{\bar{x}}^2 = (d/1.96)^2 \quad .$$

If the precision of the average grain-size estimates must be within one-half phi unit, (i.e. $d \leq .5$), then

$$\sigma_{\bar{x}}^2 \leq (.5/1.96)^2 \leq .065. \tag{3}$$

Substituting from equation (3) to equation (2) we obtain

$$.065 \geq \frac{.0291}{n_g} + \frac{.2275}{n_g n_s}$$

In fact, Kelley and McManus did not use the standard normal deviate, z, but employed the Student's t value, applicable for cases with small sample size. This led to the variance constraint $\sigma_{\bar{x}}^2 \leq 0.0404$, which also is used in the graphical solutions in Figures 2 and 3. The sensitivity of the variance to changes in t corresponding to a range of number of degrees of freedom is illustrated in Figure 4.

We can now formulate the problem in optimization terms as follows. The objective criterion is cost. The problem is to find the number of grab samples and the number of subsamples so as to minimize cost subject to a precision of no more than $\pm.5\phi$ for average grain size. In formal terms, we must

$$\text{minimize } c_t = 5n_g + 5n_g n_s$$

$$\text{subject to } 0.0404 \geq \frac{.0291}{n_g} + \frac{.2275}{n_g n_s}.$$

Because there are only two decision variables, n_g and n_s, we can solve this problem graphically. Figure 2 shows a graph with total cost, c_t, contoured as a function of n_g and n_s. This was done by simply evaluating the cost equation for a series of different values of n_g and n_s. Clearly the cost increases with an increase in n_g and n_s. The surface is nonlinear because of the product term $n_g n_s$ in the equation.

Superposed on the diagram is a line representing the nonlinear inequality constraint. The position and shape of this line was found by evaluating the variance constraint in the equation for various values of n_g and n_s. Because the constraint is an inequality, the solution for n_s and n_g is represented by the shaded area on the upper right side of the line. This is the area of feasible solution. Determining an optimal solution amounts to finding a point within the area of feasible solution at which cost is minimized.

In this example we are only interested in solution for integral values of n_g and n_s. There are three optimal solutions

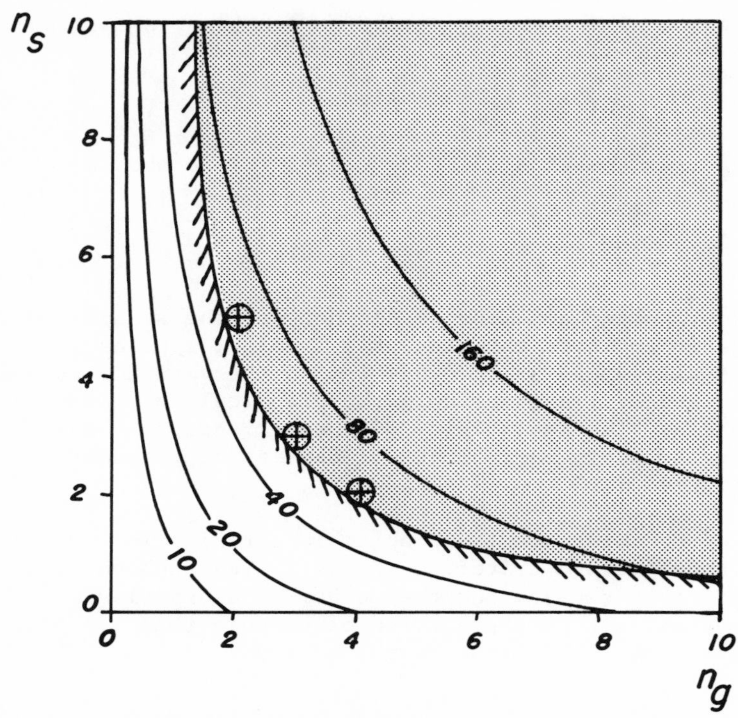

Figure 2. - Graph showing objective function, cost, contoured as function of decision variables n_g and n_s. Shaded region is area of feasible solution, bounded below by variance constraint imposed by requiring precision of ±.5ϕ for average grain size. Three equally optimal solutions lie at points shown, at integral values of n_g and n_s.

that all cost \$60 at $(n_g, n_s) = (3, 3)$, $(2, 5)$, and $(4, 2)$ as indicated on the diagram.

We also can turn the problem around, and assume that we can only spend a fixed amount of money (say \$40 per station) yet wish to make the confidence interval for average grain size as small as possible. For this situation we wish to minimize the variance of the mean, $\sigma^2_{\bar{x}}$, where

$$\sigma^2_{\bar{x}} = \frac{0.0291}{n_g} + \frac{0.2275}{n_g n_s}$$

subject to the constraint that

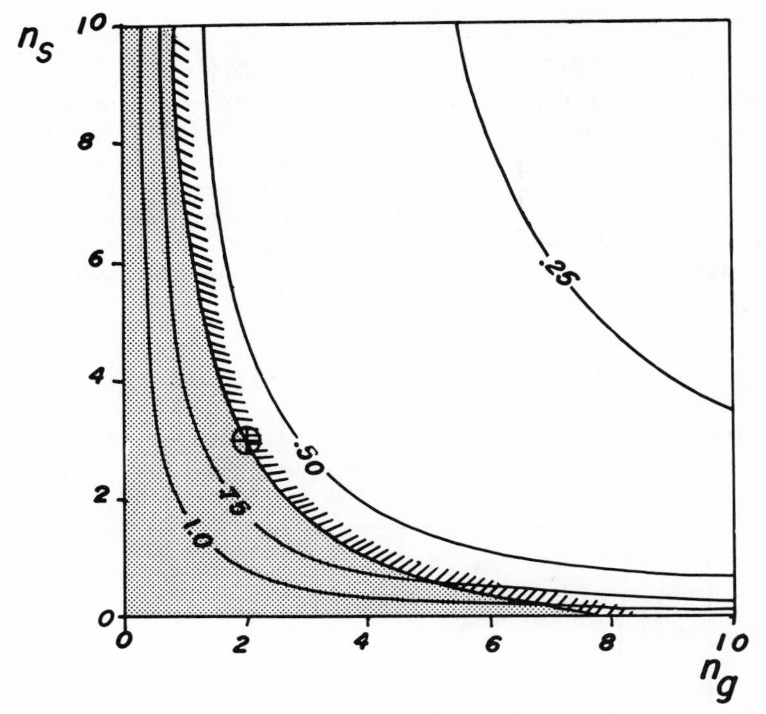

Figure 3. - Similar to Figure 2, but with objective function and constraint reversed. Here precision in phi units is contoured as function of n_g and n_s. Inequality constraint imposed by cost of \$40 bounds area of feasible solution (shaded). Precision is maximized at point shown, $n_g = 2$, $n_s = 3$.

$$40 \geq 5n_g + 5n_g n_s \quad .$$

In other works, we just reverse the objective function and the constraints.

The solution to this problem is shown graphically in Figure 3. The contours represent confidence intervals in phi units corresponding to different values of $\sigma^2_{\bar{x}}$. As the contours show, the size of the confidence interval decreases with increasing n_g and n_s. In other words, the greater the number of samples, the greater will be the precision in estimating mean grain size. The inequality corresponding to the cost constraint of \$40 defines an area of feasible solution which in this situation is to the lower left of the inequality. The graphical solution is determined by finding the point at which the confidence interval is minimized within the area of feasible solution. The optimal solution for integral values lies at $(n_g, n_s) = (2, 3)$.

In this manner objective and optimal decisions can be made regarding sampling plans. As Kelley and McManus (1969) point out, it may be that such an analysis will show that a particular study is not feasible if the precision needed costs more money than is available for the project. In a more recent paper, Kelley (1971) explores the same sampling problem, involving not only the estimation of average grain size, but also several other sedimentary parameters.

OPTIMIZATION IN DATA ANALYSIS

Many data-analysis methods used by sedimentologists and others employ built-in optimization criteria. Harbaugh and Merriam (1968) review the major data-analysis models as applied to strati-

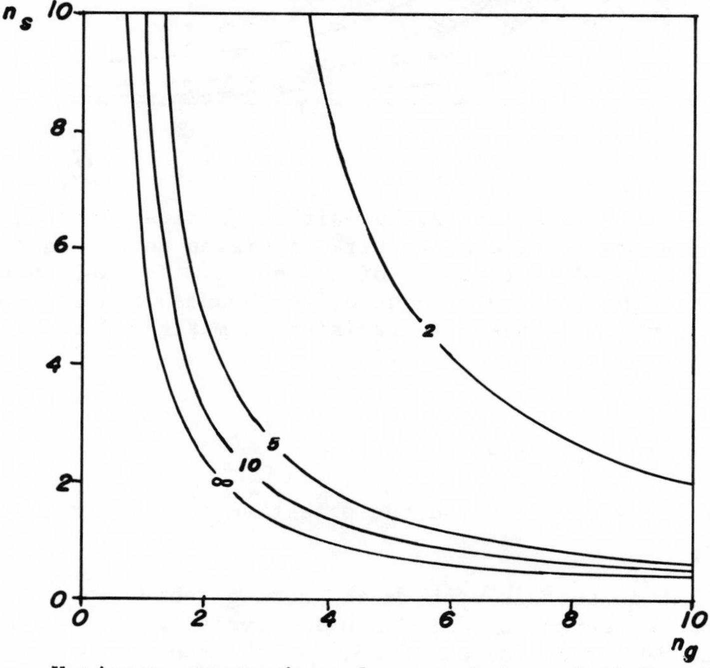

Figure 4. - Variance constraints for precision of ±0.5 phi shown as function of different values of Student's t distribution corresponding to number of degrees of freedem shown. Curve for infinite number of degrees of freedom is same as curve for standard normal deviate, z. Where number of degrees of freedom is less than 5 position of curve changes rapidly.

graphic analysis. We will briefly look at an application of regression analysis, in order to illustrate the part played by optimization.

The regression technique of least squares is widely used to fit equations of various sorts to empirically measured data points. The usefulness of the technique lies in being able to predict the behavior of a dependent variable, given the value of one or more independent variables. Simple regression of one variable on another, trend-surface analysis, and multivariate regression analysis all employ a least-squares assumption. The form of the regression equation may be a well-known mathematical function such as a polynomial or Fourier series, or may be based on terms which reflect basic physical or chemical principles applicable to the data under consideration.

Let us consider the problem of fitting a line to data points by the method of least squares. This, as the name implies, is clearly an optimization problem.

Allen (1970, p. 77) reports that field observations on the wavelength, l, of current ripples in alluvial channels vary with water depth, d, according to a power law of the form

$$l = ad^b \quad ,$$

where a and b are constants equal to 1.16 and 1.55, respectively. In order to illustrate the optimization involved in least squares, I fitted the power law equation to experimentally determined data from Guy, Simons, and Richardson (1966). These data pertain to a flume experiment using .45 mm sand, and wavelength of ripples and depth of water were recorded along with many other variables.

Let us state the problem in optimization terms. We need to determine the values of a and b (the decision variables) for which the sum of squared deviations, F, of observed wavelengths from predicted wavelengths is a minimum.

$$F = \sum_{i=1}^{n} (l_i - l_i')^2 \quad , \tag{4}$$

where F = sum of squared deviations (objective function),

l_i' = predicted wavelength for i-th data point,

n = number of data points, but

$l_i' = ad_i^b$ so substituting for l_i' into equation (4)

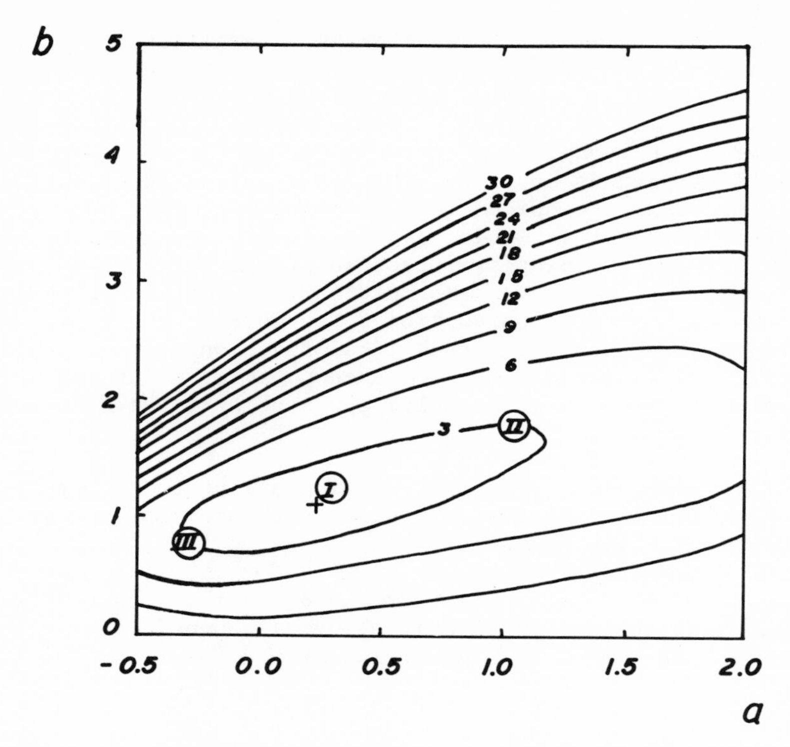

Figure 5. - Sum of squared deviations, F, contoured as function of decision variables, a and b, for power law $l = ad^b$. Optimal solution lies at point labelled I.

we obtain

$$F = \sum_{i=1}^{n} (l_i - ad_i^b)^2 .$$

One method to find those values of a and b for which F is a minimum is to calculate F for a whole range of values of the decision variables. Figure 5 shows a graph on which F has been contoured as a function of a and b. The calculations were made on a time-sharing computer terminal in just a few moments and were plotted by hand. We now can see that the objective function has a minimum bounded by the contour at $F = 3$. This area covers rather a wide range of values of a and b, a ranging from .8 to 1.8 and b from -.3 to 1.2. This implies that a wide range in regression lines will give similar fits.

We could refine our optimal values for a and b more closely by calculating more values of F close to the inferred minimum. Alternatively we can obtain an exact solution for a and b by setting $\frac{\partial F}{\partial a} = \frac{\partial F}{\partial b} = 0$, and solving the resulting equations.

On the graph, Figure 5, the exact solution is marked I. The solutions defined by the points I, II and III are shown as straight lines on a log-log plot of data points (Fig. 6). Clearly there is an enormous number of lines with different slopes and intercepts that could be fitted to these data without greatly affecting the degree of fit. The contoured surface for the objective function clearly demonstrates this, too. The graphical solution is a valuable addition to the exact calculus solution because it immediately indicates the sensitivity of the objective function to changes in the decision variables. The object here is not to champion the validity of a power law as applied to these data, but simply to illustrate the nature of the optimization process as it is found in least-squares regression.

Other data-reduction techniques such as factor analysis, discriminant-function analysis, cluster analysis, and several more, are used by sedimentologists. All these techniques involve the use of optimization criteria. McCammon (1970) describes an interesting

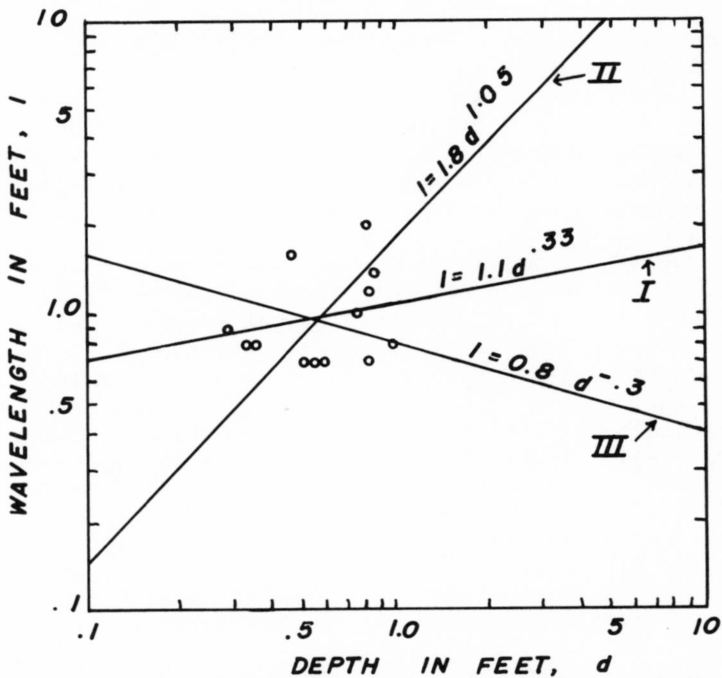

Figure 6. - Graph of log depth versus log wavelength showing data points obtained by Guy, Simons, and Richardson (1966) using .45 mm sand in series of flume experiments. Wide variety of straight lines could be fitted to points.

application of optimization to a problem of estimating proportions of minerals in sedimentary rocks.

OPTIMIZATION OF SIMULATION MODELS

Mathematical models are increasingly used for simulating sedimentary processes. In some examples the models are relatively simple, and exact solutions can be obtained by analytical means. For example, the models described by Allen (1970) are based on physical principles, employing equations relating the characteristics of terrigenous sedimentary deposits to environmental parameters such as water depth, current velocity, and so on. These models are invaluable for prediction and insight and require only a modest amount of computation for solution in particular instances.

More complex mathematical models, whose solutions are difficult to obtain by hand (either because analytical solutions do not exist or because the volume of computation is prohibitive), may be programmed for computers. They are generally known as computer-simulation models.

For example, the delta model described by Bonham-Carter and Sutherland (1968) is constructed from a large number of simple predictive equations, similar to the ones published by Allen, and linked so that a series of sedimentological events can be simulated in space and time. Deterministic models of this general type also include one for evaporite sedimentation (Briggs and Pollack, 1967), estuarine sedimentation (Farmer, 1971), deposition of volcanic ash and resulting bentonite (Slaughter and Hamil, 1970). Computer simulation models also may include stochastic components, thereby incorporating into the mechanism of the model the variability that is a predominant feature of sedimentological systems. Included in this general category are models of algal-bank growth (Harbaugh, 1966) and formation of a complex recurved spit (King and McCullagh, 1971). Also included in this category are the Markov-chain models of stratigraphic sequences; some recent publications include Lumsden (1971), Read (1969), Schwarzacher (1969) and Dacey and Krumbein (1970).

Let us now look at how optimization plays a role in mathematical-simulation models.

Role of Optimization

Optimization is an integral part of any simulation process. The parameters in the model, and even the model itself, must usually by adjusted so as to make the output from theoretical experiments accord with the real world that is being modeled. This is normally accomplished by making a succession of computer runs, altering the input variables so as to minimize the differences between experimental results and the real-world prototype.

There is another, less direct, method by which optimization plays a part in simulation modeling. This is in sensitivity analysis whereby the effects of changing one or more parameters is evaluated in terms of the sensitivity of the model. There may be no attempt to locate maxima or minima, but, nevertheless, the process of systematically changing the decision variables is identical to an optimization procedure, so one can think of sensitivity analysis as a logical extension of optimization.

We will now examine a particular application of simulation modeling in order to illustrate sensitivity analysis. The model is a simple one related to the vertical distribution of suspended-sediment discharge in an alluvial channel. It is an aspect of the Bonham-Carter and Sutherland (1968) delta model that was not explored in the earlier paper.

Suspended-Sediment Transport Model

The vertical velocity distribution for water flowing in an open channel, and the manner in which suspended sediment is distributed with depth are both well known, and equations for both relationships have been substantiated by both field and laboratory observations. The sediment discharge, or total mass of sediment passing a unit cross section of channel in a unit of time, can be obtained by integrating the product of the velocity and sediment concentration over depth. The integration is not possible to solve exactly by analytical means, but can be obtained by numerical computation.

I have found that the sediment-discharge variation with depth in an open channel is sensitive to changes in grain size of the suspended material.

An excellent discussion of sediment transport in alluvial channels and its effects on sorting is found in Brush (1965). Vanoni, Brooks, and Kennedy provide a useful summary of mathematical relationships describing flow in alluvial channels.

The two-dimensional velocity distribution for flow in an open channel is given by

$$V_z = \bar{V} + \frac{U_*}{k} (1 + ln \frac{z}{d}) ,$$

where $U_* =$ shear velocity $= \sqrt{gdS}$,

$k =$ von Karman's constant ,

$d =$ depth ,

$\overline{V} =$ mean velocity ,

$V_z =$ velocity at a distance z above the bottom ,

$S =$ slope , and

$g =$ acceleration due to gravity.

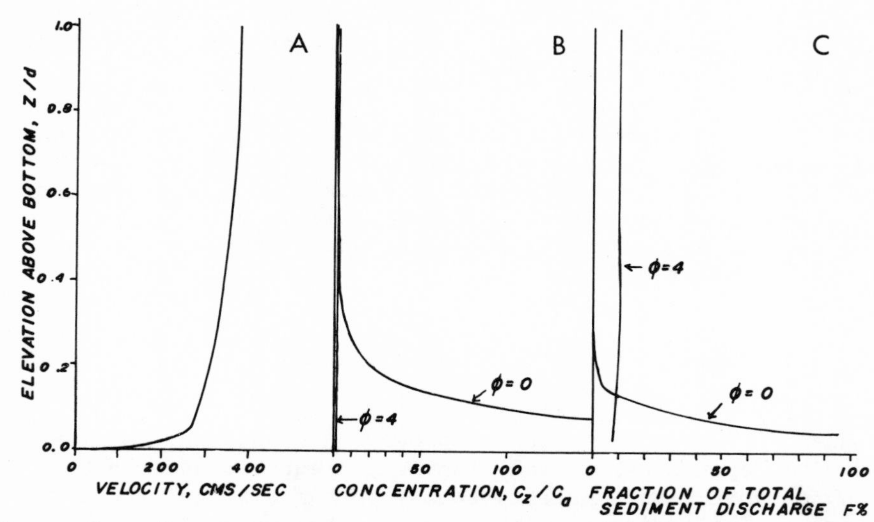

Figure 7. - Vertical profiles in two-dimensional alluvial channel. A, Velocity profile; B, suspended-sediment concentration profiles for two grain sizes; C, sediment-discharge profiles for two grain sizes. In B and C, functions are evaluated to distance of 4 cms above bottom.

Figure 7A shows the velocity profile obtained with the relationship in an open channel with a slope of .001, depth of 300 cms and a value of von Karman's constant of 0.4. The latter constant changes somewhat depending on the amount of sediment in suspension; $k =$ 0.4 is for clear water.

The sediment concentration profile is given by

$$C_z = C_a \left(\frac{d-z}{z} \cdot \frac{a}{d-a}\right)^P ,$$

where C_z = concentration at distance z above bottom,

C_a = reference concentration at distance a above bottom,

P = w/U_*k, and

w = fall velocity of sediment grains in still water.

The ratio C_z/C_a has been plotted in Figure 7B for two different grain sizes using the channel parameters described. The grains are assumed to be quartz spheres. Fall velocities were obtained using a relationship and values published by Gibbs, Matthews, and Link (1971). Notice that the material with diameter $\phi = 0$ is concentrated close to the bottom. At $\phi = 4$ the concentration is more uniform with depth.

From these concentration and velocity relationships, one may calculate the sediment discharge for different elevations above the bottom of the channel. The sediment discharging between elevations z_1 and z_2 is given by the integral

$$\int_{z_1}^{z_2} VC\, dz .$$

The fraction, F, of the total suspended load travelling between z_1 and z_2 is thus

$$F_{z_1,z_2} = \frac{\int_{z_1}^{z_2} VC\, dz}{\int_{\varepsilon}^{d} VC\, dz},$$

where ε = small elevation close to the bottom, introduced because at $z = 0$, the concentration equation goes to infinity.

After simplification, this reduces to

$$F_{z_1,z_2} = \frac{\int_{z_1}^{z_2} V(d-z/z)^P\, dz}{\int_{\varepsilon}^{d} V(d-z/z)^P\, dz} .$$

This can be readily solved by numerical integration. By subdividing the channel into ten horizontal streamtubes, the proportion of the total suspended-sediment discharge can be evaluated for each streamtube. These calculations were determined for two grain sizes and the results shown in Figure 7C.

The sensitivity of the parameter F to grain size as well as distance above the bottom of the channel was evaluated using the same settings of the hydraulic parameters as shown in Figure 7. Figure 8 shows F contoured as a function of grain size and elevation above the bottom.

For phi diameters greater than 4, the sediment discharge is uniform with depth; for phi diameters less than 0, all sediment is transported in the lowermost streamtube. Between these extremes the mode of transport undergoes a rapid change. The most rapid change occurs in the medium sand range. At $\phi = 2$, 75 percent of the discharge is transported in the upper nine streamtubes; at $\phi = 1$, however, the same streamtubes only carry 40 percent of the total discharge.

The actual size distribution of the total sediment load has not been considered here. It is assumed that each grain size is equally abundant. It would be straightforward, however, to convert these numbers for a sediment load of known grain-size dis-

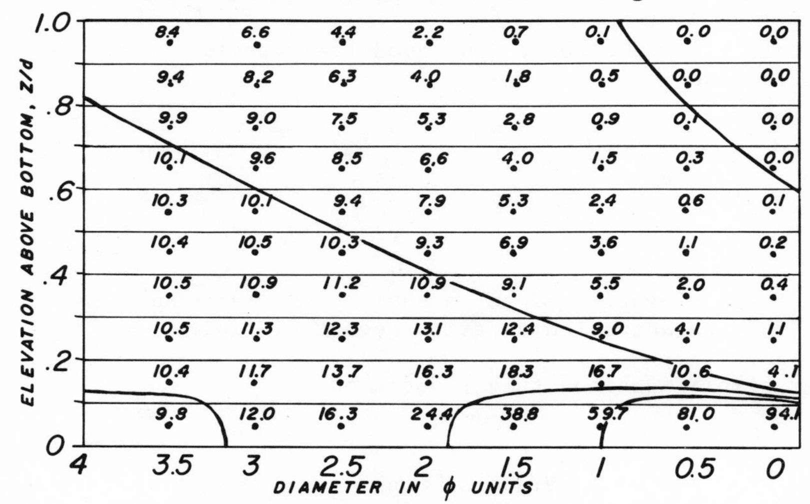

Figure 8. - Suspended-sediment discharge, F, contoured as function of grain size and elevation above bottom. Depth of channel is 300 cms, mean velocity is 343 cms per sec, shear velocity is 17.15 cms per sec. For each grain size, values of F were obtained by integrating vertically within each of ten streamtubes.

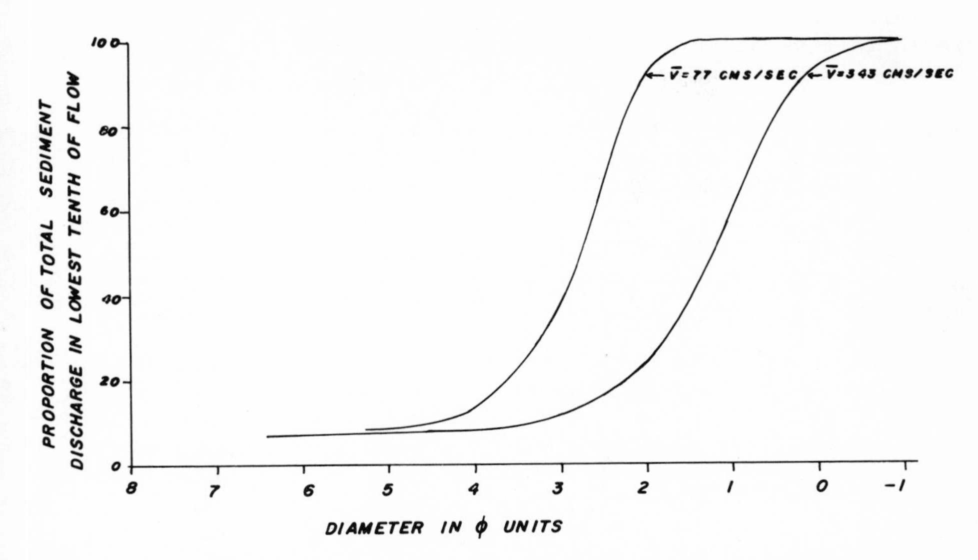

Figure 9. - Suspended-sediment discharge, F, in lowest tenth of flow (bottom streamtube) as function of grain size. Notice that as mean velocity is lowered, transition from near-bottom transport to uniform transport throughout water column becomes more abrupt as shown by steeper slope for $\overline{V}$ = 77 than for $\overline{V}$ = 343.

tribution in a similar fashion to that described by Brush (1965).

Another factor not considered is the effect of changing the hydraulic parameters of the channel. How does the model respond to a change in average velocity of the channel, for example? Figure 9 shows the effect of reducing the average current velocity, $\overline{V}$, from 343 cms per sec to 77 cms per sec (this was accomplished in the computer program by reducing the slope of the channel bottom). The figure displays the values of the discharge, F, for the lowermost streamtube only. The abrupt change from a condition in which the sediment is carried close to the bottom, to one in which the sediment is distributed uniformly throughout the water column is denoted by the slope of the line. The line for $\overline{V}$ = 77 cms per sec is steeper than the line for $\overline{V}$ =343 cms per sec indicating an even more abrupt transition for the lower mean-velocity conditions.

Interpretation of Results

What is the cause of this rather abrupt change and what is the effect of this transition on sediment deposition?

Clearly the key factor in causing the transition is the rapid change in fall velocity over this range of grain sizes.

Table 1. Table showing fall velocity and w/kU_* (exponent in concentration equation) as function of grain diameter for quartz spheres at 20°C.

Diameter, d (ϕ units)	Fall velocity, w 1/ (cms/sec)	w/kU_* 2/ (dimensionless)
5	0.1	0.015
4	0.3	0.044
3	1.1	0.160
2	3.2	0.466
1	7.6	1.108
0	15.3	2.230
-1	27.4	3.994
-2	44.9	6.545

1/ from Gibbs, Matthews, and Link (1971)
2/ $k = 0.4$, $U_* = 17.15$ cms/sec at $\overline{V} = 343$ cms/sec

The fall velocity, in turn, affects the value of $P(=w/kU_*)$, the exponent in the equation for suspended sediment concentration. Table 1 shows that, for the situation of $\overline{V} = 343$, $U_* = 17.15$, $k = 0.4$, P has a value close to 1 for phi diameter of 1. For grain sizes either side of this value, P changes rapidly from unity, and the effect on the sediment concentration term becomes pronounced. This, then is the cause of the rapid transition of sediment discharge, F, with grain size.

This rapid switch-over from a condition where essentially all the grains larger than a critical size are carried close to the bottom, to one where all the grains smaller than the critical size are almost uniformly transported throughout the water column, is borne out by observations. G.V. Middleton (personal comm., 1971) pointed out that the grain-size distribution of the wash load of streams is distinct from that of the load transported close to the bed. The abrupt difference between the two may correspond to the similar rapid transition noticed in this computer experiment.

The sediment sorting at the mouth of the delta in the Bonham-Carter and Sutherland model was undoubtedly influenced by this relationship also, although several other factors also were present, such as the size distribution of the original sediment load, depth of water through which grains could settle, and others. The coupling of this suspended-sediment discharge model with other models for deposition may lead to further insights into sediment sorting in alluvial channels.

In the present context, however, the usefulness of sensitivity analysis using a computer model is stressed. In this situation, well-known mathematical relationships, substantiated by both field and laboratory studies, were combined in a mathematical model that could not be solved exactly. The computer model permitted experimentation in a manner similar in some respects to using a flume.

CLOSING REMARKS

Perhaps one of the most challenging aspects of mathematical modeling in sedimentology is to discern optimality principles. We may find that optimization will provide an important key to unscrambling complex sedimentary patterns. As Langbein (1966) shows for river geometry, the objective functions may be complex, comprising a combination of several variables. Nevertheless, if the costs of poor design of sedimentary structures, for example, can be determined, this will lead to a clearer understanding of their formation.

Besides trying to unearth the blueprints for nature's design, the plans for data analysis and for sampling can be laid out using optimization criteria that are fundamentally analogous to those found in physics, biology, and other disciplines. By recognizing and studying methods and techniques of optimization, the analysis of sedimentary processes has much to gain.

ACKNOWLEDGMENTS

I wish to thank Walter Schwarzacher for reading the manuscript and Pat Mench for drafting the illustrations.

REFERENCES

Allen, J. R. L., 1970, Physical processes of sedimentation: American Elsevier, New York, 248 p.

Bagnold, R. A., 1956, The flow of cohesionless grains in fluids: Roy Soc. London Phil. Trans., v. 249, ser. A, p. 235-297.

Bonham-Carter, G. F., and Sutherland, A. J., 1968, Mathematical model and FORTRAN IV program for computer simulation of deltaic sedimentation: Kansas Geol. Survey, Computer Contr. 24, 56 p.

Briggs, L. I., and Pollack, H. N., 1967, Digital model of evaporite sedimentation: Science, v. 155, no. 3761, p. 453-456.

Brush, L. M., 1965, Sediment sorting in alluvial channels, in Primary sedimentary structures and their hydrodynamic interpretation: Soc. Econ. Paleon. and Mineral. Sp. Publ., no. 12, p. 25-33.

Cochran, W. G., 1963, Sampling techniques (2nd ed.): John Wiley & Sons, New York, 413 p.

Dacey, M. F., and Krumbein, W. C, 1970, Markovian models in stratigraphic analysis: Jour. Intern. Assoc. Math. Geol., v. 2, no. 2, p. 175-191.

Farmer, D. G., 1971, A computer-simulation model of sedimentation in a salt-wedge estuary: Marine Geology, v. 10, no. 2, p. 133-143.

Gibbs, R. J., Matthews, M. D., and Link, D. A., 1971, The relationship between sphere size and settling velocity: Jour. Sed. Pet., v. 41, no. 1, p. 1-7.

Goldstein, H., 1950, Classical mechanics: Addison-Wesley, Reading, Mass., p. 46-47.

Griffiths, J. C., 1967, Scientific method in analysis of sediments: McGraw-Hill Book Co., New York, 508 p.

Guy, H. P., Simons, D. B., and Richardson, E. V., 1966, Summary of alluvial-channel data from flume experiments, 1956-61: U. S. Geol. Survey Prof. Paper 462-I, 96 p.

Harbaugh, J. W., and Bonham-Carter, G. F., 1970, Computer simulation in geology: John Wiley & Sons, New York, 575 p.

Harbaugh, J. W., 1966, Mathematical simulation of marine sedimentation with IBM 7090/7094 computers: Kansas Geol. Survey, Computer Contr. 1, 52 p.

Harbaugh, J. W., and Merriam, D. F., 1968, Computer applications in stratigraphic analysis: John Wiley & Sons, New York, 282 p.

Kelley, J. C., and McManus, D. A., 1969, Optimizing sediment-sampling plans: Marine Geology, v. 7, no. 5, p. 465-471.

Kelley, J. C., 1971, Multivariate oceanographic sampling: Jour. Intern. Assoc. Math. Geol., v. 3, no. 1, p. 43-50.

King, C. A. M., and McCullagh, M. J., 1971, A simulation model of a complex recurved spit: Jour. Geology, v. 72, no. 1, p. 22-37.

Krumbein, W. C., and Graybill, F. A., 1965, An introduction to statistical models in geology: McGraw-Hill, New York, 475 p.

Langbein, W. B., 1966, River geometry: minimum variance adjustment: U. S. Geol. Survey Prof. Paper 500-C, p. C6-C11.

Leopold, L. B., and Langbein, W. B., 1966, River meanders: Scientific American, v. 214, no. 6, p. 60-70.

Lumsden, D. N., 1971, Markov-chain analysis of carbonate rocks: applications, limitations, and implications as exemplified by the Pennsylvanian System in southern Nevada: Geol. Soc. America Bull., v. 82, no. 2, p. 447-462.

McCammon, R. B., 1970, Component estimation under uncertainty, in Geostatistics: Plenum Press, New York, p. 45-61.

Read, W. A., 1969, Analysis and simulation of Namurian sediments in central Scotland using a Markov-process model: Jour. Intern. Math. Geol., v. 1, no. 2, p. 199-219.

Rosen, R. R., 1967, Optimality principles in biology: Butterworths, London, 198 p.

Schwarzacher, W., 1969, The use of Markov chains in the study of sedimentary cycles: Jour. Intern. Assoc. Math. Geol., v. 1, no. 1, p. 17-41.

Slaughter, M., and Hamil, M., 1970, Model for deposition of volcanic ash and resulting bentonite: Geol. Soc. Amer. Bull., v. 81, no. 31, p. 961-968.

Vanoni, V. A., Brooks, N. H., and Kennedy, J. F., 1961, Lecture notes on sediment transportation and channel stability: W. M. Keck Lab. for Hydraulics and Water Res., California Inst. Tech., Tech. Rept. KH-R-1.

Wilde, D. J., and Beightler, C. S., 1967, Foundations of optimization: Prentice-Hall, Englewood Cliffs, N. J., 480 p.

INTERPRETATION OF COMPLEX LITHOLOGIC SUCCESSIONS BY SUBSTITUTABILITY ANALYSIS

John C. Davis and J. M. Cocke

Kansas Geological Survey and

East Tennessee State University

ABSTRACT

Many stratigraphic successions are characterized by repetitive patterns of lithologies. These patterns are most apparent if lithologies are grouped into relatively few categories, and become increasingly obscure as rock types are classified into finer subdivisions. Most cyclothems and megacyclothems, for example, are patterns composed of only four or five distinctive lithologies. Unfortunately, the gross classification necessary to reveal a cyclic pattern results in lithologic categories which yield meager environmental information.

A section through supposedly cyclic lower Pennsylvanian rocks in eastern Kansas was examined and the lithologies classified into 17 states. Although this degree of subclassification is typical of lithofacies studies, the variety of rock types conceals any cyclicity that might be present. Seemingly different lithologies appear at common positions within cyclothems, obscuring the repetitive pattern in the sequence. These lithologies "substitute" for one another in successive cycles, but may be identified by substitutability analysis, a classification procedure that groups states on the basis of their context in a sequence. States with common high conditional probabilities on subjacent and superjacent states are considered equivalent. Results suggest that lithologies must be combined into fewer than eight states before a cyclic pattern emerges. Analyses also suggest that the lower Pennsylvanian cyclothems studied represent interaction of two depositional processes rather than a single megacyclic process.

INTRODUCTION

Sedimentary deposits exhibit differing degrees of perfection of a cyclic pattern. Varved deposits, for example, consist of a precise alternation between two seasonally controlled states. Certain flysch deposits exhibit an almost perfect repetitive succession of upward fining units. Some lithologies are so intimately coupled in their origins that they rarely occur out of sequence. Examples include the couplet formed by seatearth and coal. Formation of one implies the formation of the other; whereas seatearths are found without overlying coals, and coals found without underlying seatearths, the two rarely occur in reverse order. Commonly, the sequence of evaporite lithologies, limestone → dolomite → anhydrite → salt is less well developed. Chemical considerations dictate a fixed order to the succession; observed variations from the predicted order require explanations involving other geologic mechanisms.

Beyond these relatively simple examples are more complex types of sedimentary alternations. Included are the cyclothems and megacyclothems described by many authors (e.g., Merriam, 1964; Duff, Hallam, and Walton, 1967; Weller, 1960). In general, these terms refer to sequences which consist of a more-or-less regular recurrence of a variety of lithic types vertically within the stratigraphic section. Because lengthy sections rarely repeat in a precise manner, workers may refer to an "idealized cyclothem" (Weller, 1930; Weller and Wanless, 1939; Moore, 1936) which is expressed as a series of naturally occurring approximations called "typical cyclothems." The ideal cyclothem is what would develop, presumably, if the cyclothem-causing mechanism functioned without interference from other depositional controls. Its influence is impressed onto actual sections and appears as the common elements of typical cyclothems; the degree of their correspondence to the ideal reflects the relative influences of the cyclothem mechanism and other controls.

The example presented in this paper will be concerned with cyclothems and megacyclothems developed in a segment of Upper Carboniferous strata in the American Midcontinent. These deposits consist of complex alternations of a wide variety of lithic types, almost all marine in origin. No entirely satisfactory explanation has been advanced for the origin of the cyclic aspect of these deposits, although some authors (Moore and Merriam, 1965) strongly imply that the cycles resulted from widespread marine transgression and regression.

The cyclothem mechanism is a physical process, which leaves its imprint on the lithologies and faunal content of the sediment. A detailed examination of the lithologic succession should provide information which could be analyzed quantitatively to estimate the relative influence of the cyclic process. Quantitative studies of the nature of the cyclic pattern in these rocks have been made (Pearn, 1964; Schwarzacher, 1967, 1969) but these deal with idealized representations of the succession and involve simplistic definitions of lithologies. In this paper, we will attempt to measure quantitatively the cyclic aspect of these rocks, using actual data typical of that gathered in a modern stratigraphic study. If a cyclic pattern does not emerge from the analysis, the lithologic data will be iteratively generalized until a pattern becomes apparent. The degree to which lithologies must be generalized should provide information not only on the nature of the cyclothem process, but also on the relative equivalencies of rock types. Although consideration is confined to one stratigraphic interval, the methods employed should be applicable to other successions as well.

DETECTION OF CYCLICITY

A stratigraphic succession may be considered as having two principal attributes: lithology (including biotic composition) and thickness. The range of possible lithologies in sedimentary rocks is finite, but an infinite number of subdivisions are possible within this range. Meaningful, consistent identifications of rock lithologies must be made in order to recognize repetitions. On a sufficiently fine scale of classification, no repetitions are possible because every unit will be placed in an unique lithologic category. At the other extreme, an entire stratigraphic section may be placed into only one or two gross lithologic groups. Stratigraphic units in a cyclothem, or any system of repetitive lithologies, must be classified into states which are rigorously and consistently defined and which are mutually exclusive. Stratigraphic position and lithologic classification must be completely independent if any assignment of units to states is to have meaning. Assignment of a unit to a lithology on the basis of its position within a cyclothem introduces an element of circular reasoning into the definition of the cyclothem (cf. Schwarzacher, 1969, p. 29).

Cycles within a succession of lithologic states may be obscure for a variety of reasons. They may be incompletely developed, in which situation the cyclic pattern must be recognized from a succession of fragments of cycles. In the extreme, it may be impossible to deduce the nature of the ideal cycle or even to determine if cycles are present at all.

Commonly, comparatively few cycles of the magnitude of cyclothems or megacyclothems are present in any stratigraphic interval. If the cycles are incomplete, there may be insufficient repetition to verify statistically their existence. This is a severe problem when working with actual, as opposed to idealized, stratigraphic sections. Nonstationarity of the cyclic pattern presents similar problems. Examination of the Upper Paleozoic succession in Kansas, for example, shows pronounced variation in the nature of lithic repetitions from the Middle Pennsylvanian into the Lower Permian.

Inconsistency in the assignment of rock units to lithologic classes can obscure or bias the pattern of a succession. This may result from assignment of units of a single lithic type to several classes, from the inclusion of units of dissimilar lithologies into a single class, and from the generalization of a heterogeneous interval into a single unit.

Every practical precaution was taken in this experiment to attain consistency of lithologic identification and to maintain objectivity in measurement and analysis. The data were gathered on a stratigraphic section representing most of the Missourian Stage of the Pennsylvanian System, which is well exposed in quarries near Kansas City, Missouri, and along adjacent highways. The Missourian section was selected because it contains "typical" cyclothems, less perfectly developed than those in the overlying Virgilian Stage but far better than many that have been recognized elsewhere. A composite section 342 feet long was obtained, with observations taken at 6-inch intervals. The section does not contain any covered or missing intervals, nor are there any discernible lateral changes in facies between segments of the column.

All segments of the composite section, along with alternate segments, were examined prior to measurement; at that time, decisions were made as to the lithologic states present. These states are based primarily on lithology, although paleontologic content and fabric are important in the definition of some. Seventeen distinct lithologies were recognized; most samples can easily be catagorized into a single state although some ambiguity is present with certain intermediate samples. Limestone lithologies were defined by a carbonate petrographer; clastic rock lithologies were defined by a petrologist specializing in shales. During measurement of the sections, the status of ambiguous samples was decided mutually. Final data consist of 684 ordered descriptions and classifications. Because the sampling interval is constant, thickness of units may be estimated to the nearest one-half foot. However, in this study only the succession of changes in lithology are considered.

During field measurement, no attempt was made to recognize either cyclic patterns or the formal boundaries of rock units. In this manner it was hoped to avoid any tendency to assign units to "expected" classes in the cyclothem model.

Techniques for quantitative assessment of cyclicity in sedimentary successions are based on time-series analysis (Schwarzacher, 1964) or on determination of the conditional probabilities between successive states (Vistelius and Feigel'son, 1965; Krumbein, 1967; Krumbein and Dacey, 1969). No exact tests for the presence of a cyclic component are available, because stratigraphic sections almost invariably violate one or more of the fundamental assumptions required for these tests. A serious constraint is imposed by the nature of most stratigraphic data; the states are nominal classes, so many of the powerful techniques of parametric statistics are not available. Some attempts have been made to apply Fourier analysis or autocorrelation methods, but these necessitate an arbitrary scaling of states (Mann, 1967; Carss, 1967) which has an unassessed effect on the analysis. Some workers have used time-series methods to investigate variations in thicknesses of beds (Anderson and Koopmans, 1963), electrical properties (Preston and Henderson, 1964), or composition (Anderson, 1967) with somewhat more assurance.

Autoassociation (Sackin and Merriam, 1969) is a technique similar to autocorrelation, but based on the counting of common nominal states in two matched sequences. Assumptions are much less restrictive than those of autocorrelation, and the method seems appropriate for the examination of sedimentary successions. If cycles exist in a series of states, they will appear as significant high autoassociations at repeating intervals whose length is equal to the period of the cycles.

The transition probability matrix of a sequence is a matrix of conditional probabilities, and will exhibit no cyclic structure if the sequence is random. The hypothetical section shown in Figure 1 was generated from a random number table and produces a transition probability matrix in which all rows tend to be identical and the columns reflect relative abundances of the various states (in the example, all states are equally abundant). In constrast, Figure 2 shows a purely deterministic succession. The transition probability matrix is completely structured and cyclic. Intermediate between these two extremes are sequences which show partial dependency or conditional probability of one state or another. Such sequences exhibit Markovian behavior, and although pronounced conditional behavior does not necessarily imply cyclicity, cyclic successions must be Markovian.

The structure of transition probability matrices may be shown by plotting them as directed networks or flow graphs (Berge and Ghouila-Houri, 1965), considering the transition probabilities from one state to another to be vectors whose magnitude reflects the probability of one state succeeding another. A purely deterministic succession such as Figure 2 will produce a flow graph in the form of a ring. In contrast, the random succession of Figure 1 will result in a complex flow pattern in which a preferred path cannot

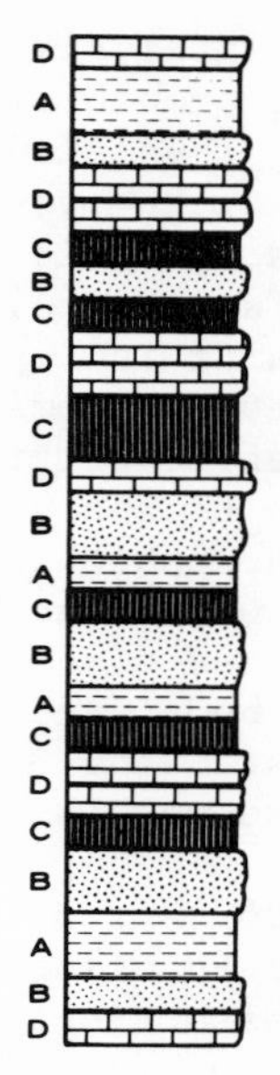

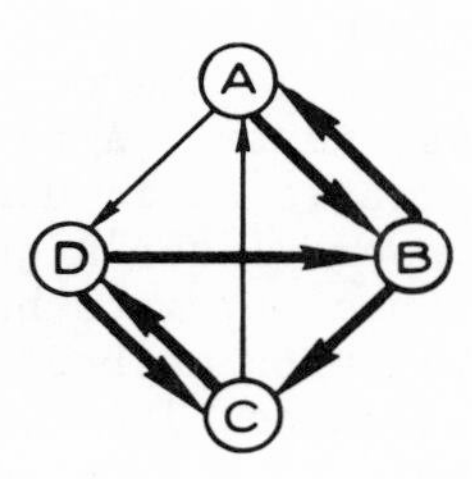

Figure 1. - Hypothetical stratigraphic section containing four lithologies, generated from random number table. Flow graph for succession does not contain preferred path.

be found.

Anderson and Goodman (1957) describe a test of the Markov property, and this has been programmed by Krumbein (1967). Unfortunately, the necessary assumptions severely restrict the applicability of this test. Because no single definitive test for the presence of cyclicity is available, several of the methods described were used concurrently in the search for cyclothems in the test section. These include examining the autoassociation of the sequences, testing of transition probability matrices for the Markov property, and examination of flow graphs for the presence of pronounced rings.

SELECTION OF LITHOLOGIC STATES

The stratigraphic interval studied has been classified by Moore (1949) into fourteen cyclothems and six megacyclothems. He recognized ten units in an ideal cyclothem, of which seven are lithologically distinct (Moore, 1936). In a later analysis, Pearn (1964) reduced these to five distinct states, and Schwarzacher (1967) further reduced the number of states to four in an attempt to measure the period of a cycle. From the experiences of other

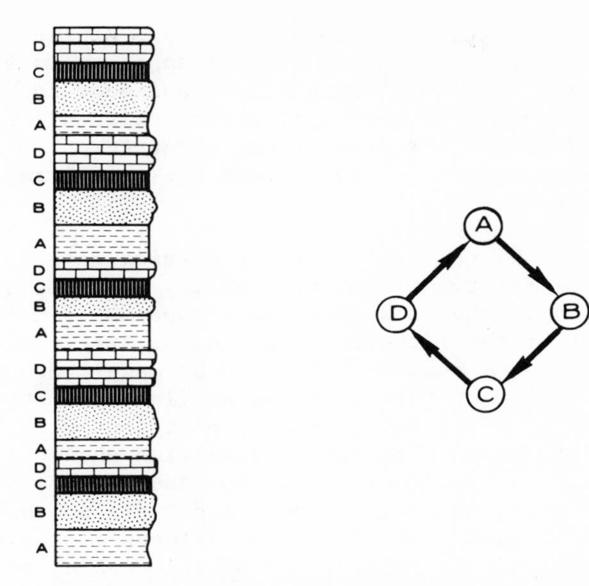

Figure 2. - Hypothetical stratigraphic section which is entirely deterministic. Resulting flow graph consists of single ring.

workers, it seems likely that some of the lithologic classes which we have recognized in the stratigraphic succession must be combined and generalized before cyclicity will become apparent.

A lithologic succession may be simplified in a number of ways. Lithologic states, judged on some basis to be similar, may be combined into single composite states. These composite states may be combined further, resulting in a reduced number of states having gross characteristics. The process of combination follows a tree-like path, consisting of successive estimates of relative similarity between states, as in cluster analysis. Estimates of similarity may be based on intuition, measurements of lithologic variables, assessment of similarities of placement within the succession, or on some other criterion.

Alternatively, stratigraphic units which contain more than one lithology may be considered to consist only of the dominant lithology. For example, thick shale units may contain sandstone or limestone beds; these would be ignored using this approach. Similarly, the complex alternation of carbonate rock types common in thick limestones would be reduced to a single lithology. Although not explicitly stated, this method of simplification seems to be used in most quantitative studies of sedimentary successions (e.g., Carr and others, 1966; Krumbein, 1967). A disadvantage of this technique is that two complex stratigraphic units containing the same collection of lithic types may be assigned to different states if the relative proportions of the constituent lithologies are different. This may actually defeat the attempt to reduce the number of states being considered, unless the classification scheme is extremely rudimentary.

Simplification of the lithologic succession results in a reduction in rank of the transition probability matrix and makes incipient patterns in the flow graph more apparent. Examinations of the flow graph or the lithologic succession itself might suggest several combinations of states that could be made. However, we will consider a method of assessing the similarity of two states by their tendency to occur in equivalent positions within the sequence. The hypothetical stratigraphic section shown in Figure 3 is characterized by unusual numbers of successions A → B → C and A → D → C. This suggests that states B and D are somehow alike, as they occur in a common context; i.e., between states A and C. States B and D are said to be <u>mutually substitutable</u>, as either may take the place of the other without altering the succession in any other way. It should be possible to measure the degree of mutual substitutability between all states in a sequence, and identify any which form natural groupings. We would expect that groups of

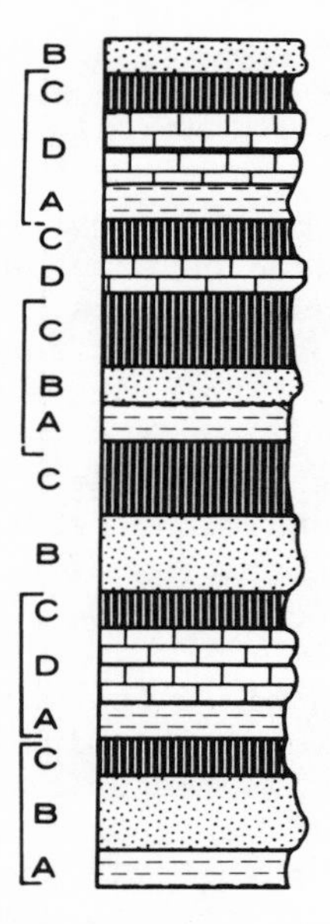

Figure 3. - Hypothetical stratigraphic section containing four lithologies arranged so unusual numbers of successions A → B → C and A → D → C appear.

mutually substitutable lithologies would correspond to those selected on the basis of geologic considerations, if the lithologic succession reflects a cyclic mechanism.

MEASURES OF SUBSTITUTABILITY

Following the development of Rosenfeld, Huang, and Schneider (1968) we can regard the sedimentation mechanism as being capable of producing any of the states $a_1, a_2, \ldots a_n$. Occurrence of any state is conditional upon the occurrence of all other states. The sequence may be denoted ΣS, whose elements are

$$\{a_{i1}, \ldots a_{ik} \mid \text{each } a_{ij} \in \{a_1, \ldots a_n\};\ k \text{ an integer} \geq 0\}.$$

For all states α, β, γ in the sequence ΣS, let $f_{\alpha\beta}(\gamma)$ be the probability that, given any segment $\alpha \rightarrow \xi \rightarrow \beta$, we have $\xi = \gamma$. We may define a function g such that

$$g(\xi,n) = \Sigma f_{\alpha,\beta}(\xi)\ f_{\alpha,\beta}(\eta)\ /\ \sqrt{\Sigma f_{\alpha,\beta}(\xi)^2\ \Sigma f_{\alpha,\beta}(\eta)^2}$$

Because of the Schwarz inequality, g will assume values only in the range 0, 1. The function $g(\xi,\eta)$ is the mutual substitutability of ξ and η is simply the normalized cross-correlation of $f_{\alpha,\beta}(\xi)$ and $f_{\alpha,\beta}(\eta)$, regarded as functions of the parameters α and β. The cross-correlation will be high only if $f_{\alpha,\beta}(\xi)$ and $f_{\alpha,\beta}(\eta)$ may be simultaneously high. This occurs only if both states ξ and η occur with high probability in many common contexts $\alpha \rightarrow \square \rightarrow \beta$.

Direct application of this concept requires knowledge of the infinite set of conditional probabilities between all elements in the sequence. For practical purposes, with finite sequence, higher order conditional probabilities can be assumed zero and only the first-order conditional probabilities considered. We are concerned then only with the set of conditional probabilities $Pr_{\alpha\rightarrow\gamma\rightarrow\beta}$ for all states γ and all combinations of α and β. This requires specification of n^3 conditional probabilities, where n is the number of states. Rosenfeld, Huang, and Schneider (1968) suggest that prior and posterior first-order conditional probabilities be considered separately, reducing the number of necessary conditional probabilities to n^2. Let $Pr_{i\rightarrow j}$ be the conditional probability that state a_j follows a_i. The first-order left substitutability of states a_r and a_s is defined as

$$L_{rs} = \Sigma Pr_{i\rightarrow r}\ Pr_{i\rightarrow s}\ /\ \sqrt{\Sigma Pr_{i\rightarrow r}^2\ \Sigma Pr_{i\rightarrow s}^2}$$

The first-order right substitutability may similarly be defined

$$R_{rs} = Pr_{r\to j} \cdot Pr_{s\to j} / \sqrt{\Sigma Pr^2_{r\to j} \cdot \Sigma Pr^2_{s\to j}}$$

Because the probability $Pr_{i\to r\to j} = Pr_{i\to r} \cdot Pr_{r\to j}$, we can create a measure of mutual substitutability between states a_r and a_s by

$$M_{rs} = L_{rs} \cdot R_{rs}$$

This procedure may be used to classify lithologic states according to their tendency to occur between common pairs of states. It is necessary to compute upward and downward transition probability matrices for the sequence of rocks and then compute two symmetrical matrices of right and left substitutability. A product matrix which contains mutual substitutabilities can then be formed by element-by-element multiplication of the two substitutability matrices. The matrices can be clustered by one of many techniques designed to express measures of relative similarity. In this study, an unweighted pair-group method suggested by Sokal and Sneath (1963) was used. This algorithm is incorporated in the NTSYS computer package implemented at The University of Kansas.

ITERATIVE COMBINATION OF STATES

An iterative approach was used to analyze the test section for cycles as the number of lithologic states is reduced. First, the plot of autoassociation versus lag was examined for significant matches, the matrix of transition probabilities tested for the first-order Markov property, and the flow diagram examined for patterns. Next, right-, left-, and mutual substitutabilities were calculated for all pairs of states, and clusters of higher mutual substitutability determined. Clusters of states then were merged to create new, composite states and the examination process repeated. Eventually, the process results in a reduced set of composite states whose succession exhibits a cyclic property. The composition of the composite states and the degree of reduction necessary to make cyclicity evident reflects the importance of the cyclothem mechanism in the stratigraphic section. If cyclothems are expressed as relatively regular patterns of lithologic successions, the pattern will emerge rapidly from the analysis. A comparatively large number of composite states will be retained, most of which will contain a limited number of lithologies which are combined in a geologically meaningful manner. A few composite states may consist of a heterogeneous collection of lithologies that occur infrequently in the original sequence. These may be regarded as lithologic "accidents" in an otherwise orderly cyclic succession.

However, if a cyclic pattern does not actually exist in the original succession, many iterations of the reduction process will be necessary before a repetitive pattern emerges. That a cyclic pattern eventually must emerge should be apparent, because the last possible iteration will result in two composite states which must of necessity oscillate. Measures of mutual substitutability will be low and the resulting composite states will be heterogeneous collections of lithologies that make no geologic sense.

APPLICATION TO MISSOURIAN CYCLOTHEMS

The stratigraphic section examined is shown in Figure 4, adapted from Zeller (1968). This section has been classified by

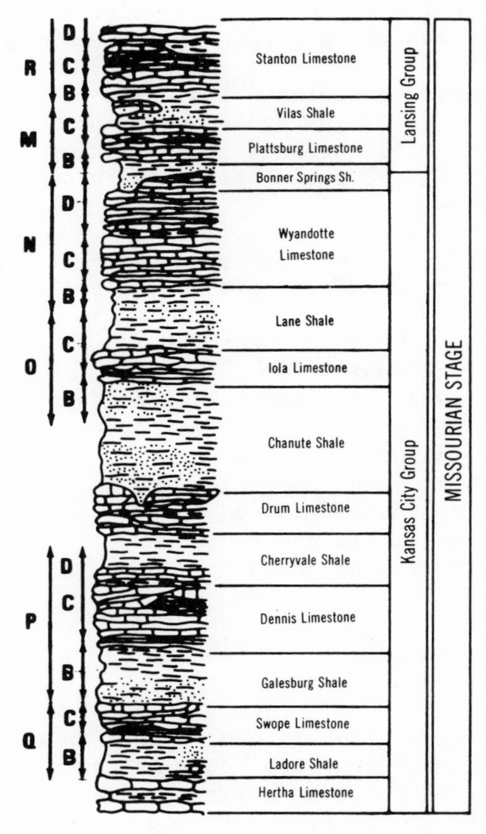

Figure 4. - Generalized stratigraphic section showing major lithologic units in interval studied. Symbols B through D represent cyclothems and M through R represent megacyclothems within interval as defined by Moore (1949) (after Zeller, 1968).

Moore (1949) into a series of cyclothems and megacyclothems which are indicated on the figure. Moore recognized three types of cyclothems which comprise the larger system of megacyclothems. His idealized "B"-type cyclothem consists of a basal sandstone unit overlain by shale which may contain coal, overlain by a thin, dense and massive limestone. An idealized "C"-type cyclothem contains, in upward succession, a basal black shale, a thin light-colored shale, capped by a thick, even-textured limestone which contains fusulinids near the bottom and algae or oolites in the top. The limestone is overlain by unfossiliferous shale. Moore's "D"-type cyclothem is not recognized on the basis of distinctive lithologies. Moore (1949, p. 82) states "Wherever recognized, ("D"-type cyclothems) are clearly separable from subjacent and superjacent cyclothems by deposits which denote retreat of marine waters, whereas the central part of such cyclothems records more or less off-shore sedimentation during marine submergence."

The measured section contains units which have been assigned to 6 B-type cyclothems, 6 C-type cyclothems, and 2 D-types. These are grouped into 6 megacyclothems which consist of either a B cyclothem followed by a C-type cyclothem, or by a succession of B-, C-, and D-type cyclothems.

Moore (1936) recognized ten major units in a ideal cyclothem, comprised of seven distinct lithologies. The cycle which he recognized is

9. Shale (and coal)
8. Shale, typically with molluscan fauna
7. Limestone, algal, molluscan fauna
6. Shale, molluscoid fauna
5. Limestone, with fusulinids and molluscoids
4. Shale, molluscoid fauna
3. Limestone, molluscan fauna
2. Shale, molluscan fauna
1c. Coal
1b. Underclay
1a. Shale with land-plant fossils
0. Sandstone

Some of these lithologies are not present in the particular section measured, although they may occur in equivalent strata at other locations. For example, no coals or underclays were encountered. Sandstones of the type referred to by Moore are likewise absent. Although Moore (1949) gives descriptions of the lithology of units in this interval, there is no exact correspondence between Moore's states and those chosen for this study. Moore describes the general characteristics of relatively large stratigraphic units over a broad area. Ours are detailed lithologic classifications at small intervals along one specific line of section. The states recognized in this study are:

B. Siltstone and very fine sandstone, thin bedded, micaceous and carbonaceous
C. Silty shales with burrows, micaceous and carbonaceous
D. Shale, brown, slightly silty, unfossiliferous
E. Shale, brown to grey, burrowed, with marine fossils
F. Shale, mottled red and green
G. Shale, green, with calcilutite cobbles
H. Shale, black, blocky and fossiliferous
I. Shale, black, papery and phosphatic
J. "Marl," fine-grained and earthy
K. Calcilutite, mottled, unfossiliferous
L. Calcilutite, laminated, with chert
M. Calcilutite, algal and sparry algal
N. Calcilutite, spar blebbed, poorly fossiliferous
O. Calcilutite, highly fossiliferous
P. Calcarenite, skeletal
Q. Calcarenite, pelletal
R. Calcarenite, oolitic skeletal

It is obvious that these states are subdivisions of a wide spectrum of lithic types. Further subdivision is possible, and probably would be made if a detailed petrographic study were undertaken of these sediments. The nine clastic lithologies contain some states which are distinguished by relatively subtle differences. States B, C, and D, for example, are differentiated by their relative amounts of sand and silt. Similar gradations exist between some carbonate states such as O and P, which are classified on the basis of the relative amount of skeletal material. Although samples with intermediate characteristics lead to ambiguous classifications, most can be consistently assigned to these seventeen categories.

The initial sequence of seventeen recurring lithic states shows little tendency toward cyclic behavior. The flow graph (Fig. 5) contains several oscillatory pairs but only one ring, involving a calcilutite, a calcarenite, and a shale. Shale and limestone are distinctly separate in the diagram. If a succession were generated from this flow graph, it would quickly evolve into a series of alternations between fossiliferous calcilutite and about three other limestone lithologies, plus an occasional fossiliferous marine shale. Alternatively, it might evolve into a series of oscillations between fine sandstones and silty shales. (These lithologies, which commonly form thick sequences in the Pennsylvanian succession of Kansas, are called "outside shales" because they separate limestone formations. Other shales, which occur as thin beds within limestone formations, are referred to as "inside

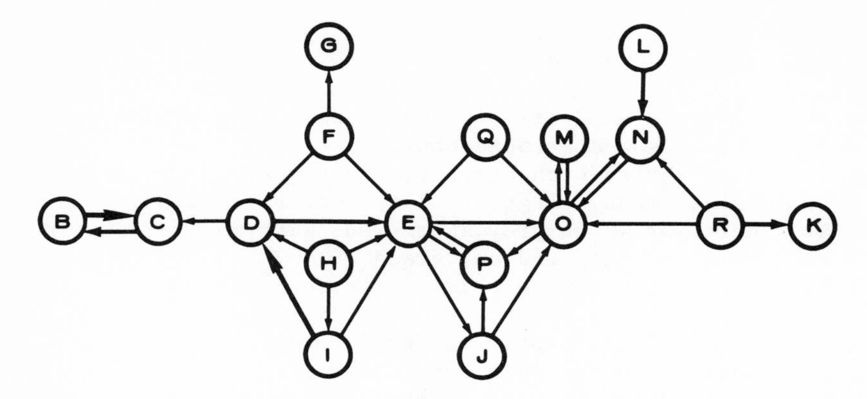

Figure 5. - Flow graph representing transitions between seventeen lithic states in interval studied. Note general absence of rings.

shales" (Schwarzacher, 1969).) The probability of a transition between these two end states is small.

The autoassociation plot (Fig. 6) reflects the pronounced tendency for oscillation between pairs of states which is apparent in the flow diagram. At least three oscillations tend to occur in succession; positive autoassociations at lags of 25 to 32 also reflect oscillatory sets of lithologies. Only three states are involved to a significant extent in producing high matches. These are algal calcilutite, fossiliferous calcilutite, and skeletal calcarenite.

The original sequence was reduced by successively combining states which have high mutual substitutabilities. During initial iterations, four pairs of lithologies, all carbonates, were combined. In the final iterations, all of these combined to form a single category containing all limestones except algal calcilutite. By the ninth iteration, the system was reduced to ten lithologies and all remaining mutual substitutabilities were below 0.3. The flow diagram is shown in Figure 7. Note that shales were relatively unaffected by the reduction process. Limestone lithic types are similar in that they have a high probability of alternating with algal calcilutite. Shales, in contrast, show low preferred transitions into other states and hence have low mutual substitutabilities.

The autoassociation plot (Fig. 8) shows a pronounced tendency to oscillate. This is caused by interbedding of various shale lithologies and limestone, and by limestone-algal calcilutite alternations. Up to six successive oscillations tend to occur

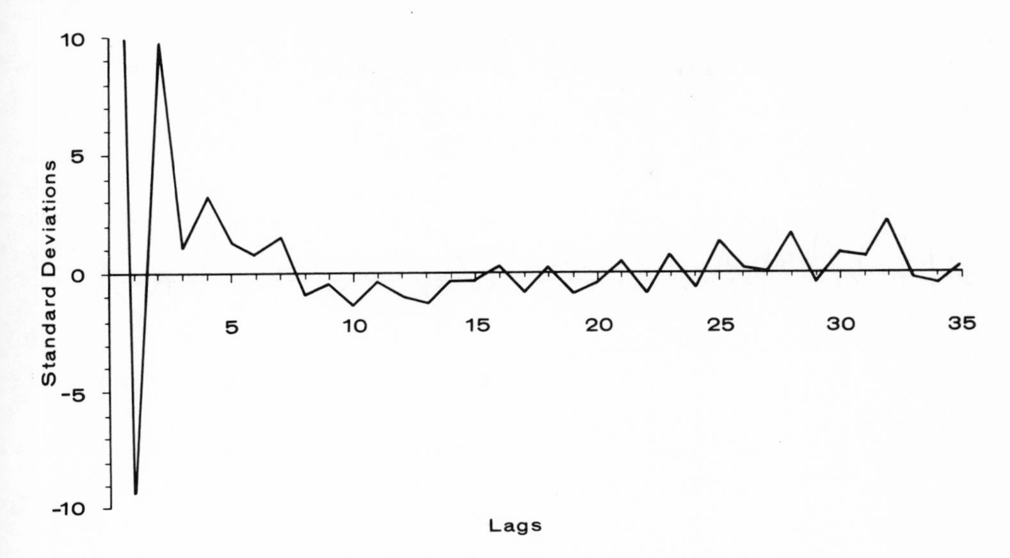

Figure 6. - Autoassociation plot of stratigraphic succession coded into seventeen lithic states. Abscissa is given in lags, ordinate in units of standard deviation from binomial mean. Binomial mean is calculated as expected number of matches of random sequence with itself.

together. High autoassociations at about lag 30 result almost entirely from matches between limestone units. High autoassociations also occur at about lag 70, and also result from matches between limestones. The occurrence of beds of each state in the sequence is shown diagrammatically in Figure 9.

Although the mutual substitutability criterion succeeded in reducing the number of lithic states, the resulting succession

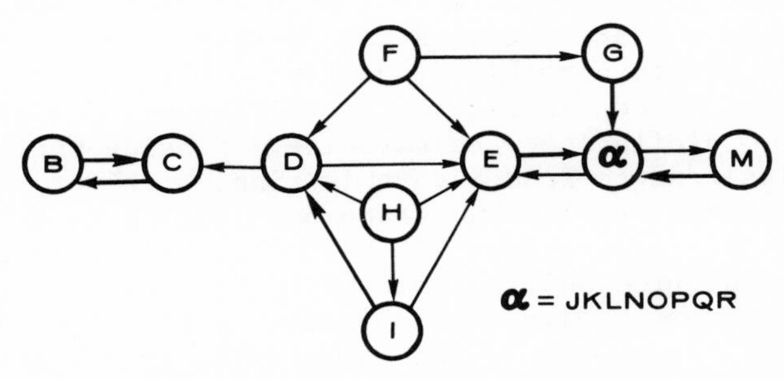

Figure 7. - Flow graph of stratigraphic section after combining states having high mutual substitutabilities.

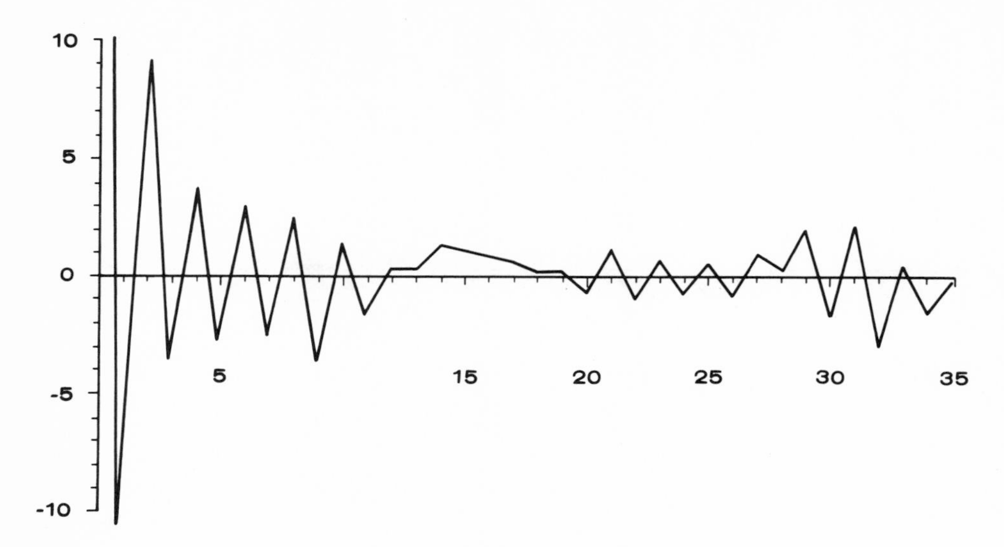

Figure 8. - Autoassociation plot of stratigraphic section after combination of states by mutual substitutability. Note pronounced oscillation caused by interbedded shales and limestones.

does not display the cyclic pattern that might be expected. However, right substitutabilities between lithic states tend to be much higher than left substitutabilities. As right substitutability is a function of upward transition probabilities, it seems reasonable that it might provide a more meaningful measure of similarity than mutual substitutability, which is also dependent upon downward transitions.

From the initial sequence, a series of six iterations reduced the maximum right substitutability below 0.5 and resulted in eight lithologic states. The composite states include an "outside shale", an "inside shale", a black shale, a calcilutite, and a calcarenite. Two other shales remain uncombined, as does oolitic calcarenite. The flow diagram is shown in Figure 10. A pronounced ring exists, "inside shale" → calcilutite → calcarenite → "inside shale". All other lithologies tend to feed into this ring, with very diffuse return. The tendency toward two-state oscillation is reduced (Fig. 11), probably as a result of the relative importance of the three-component ring and the combination of states B and C into a single category. High autoassociations at about 25 lags result from repetitious occurrences of calcilutite → calcarenite oscillations separated by intervals of calcilutite → calcarenite → shale successions. Using the symbols on Figure 10, the sequence first oscillates $\alpha \rightarrow \Psi \rightarrow \alpha \rightarrow \Psi \ldots$, then $\gamma \rightarrow \alpha \rightarrow \Psi \rightarrow \gamma \rightarrow \alpha \rightarrow \Psi \ldots$

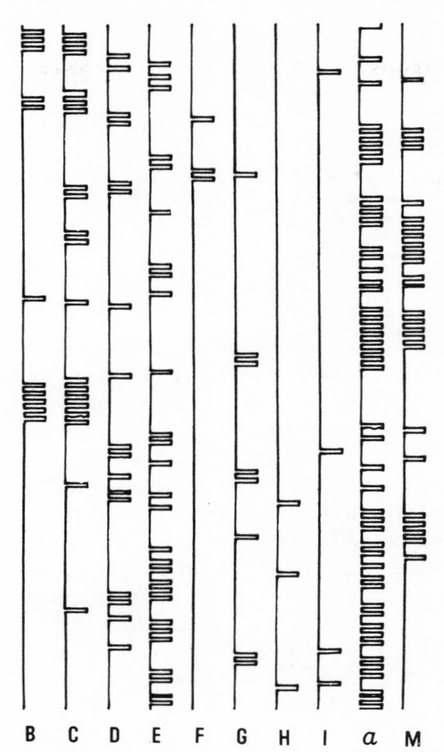

Figure 9. - Diagrammatic section showing occurrence of lithic types within interval measured. Deflection to right in line represents occurrence of that lithology. See text and Figure 9 for key to lithologies.

Within about 15 lags, the sequence again becomes $\alpha \rightarrow \Psi \rightarrow \alpha \rightarrow \Psi \ldots$, giving the longer period phenomenon seen in Figure 11. The other lithologies occur in the sequence, but not in an apparent pattern. The sequence of states is shown in Figure 12.

The iteration procedure was continued for six additional cycles, reducing the sequence four states, an "outside shale", an "inside shale", a calcarenite, and a calcilutite. The flow graph (Fig. 13) shows two rings, "inside shale" → calcilutite → calcarenite, and the reverse. "Outside shales" may enter the pattern at any point, but have low return probabilities. The maximum right substitutability after twelve iterations is less than 0.4, between inside and outside shales.

Figure 14 is a plot of autoassociation and shows the influence

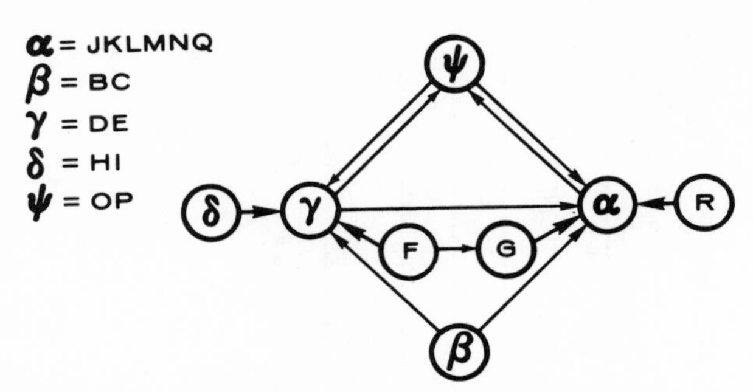

Figure 10. - Flow graph for stratigraphic section after states have been reduced by right substitutability. Diagram shows system after six iterations.

of both oscillations between pairs of states and transitions between three states. High autoassociations at initial lags result from alternations between calcilutite and calcarenite and between calcarenite and "inside shale". High autoassociations at lags 14 and 18 result from repetitions of the two limestone types. High autoassociations at about lag 30 reflect changes in the repetitive pattern from calcilutite → calcarenite → calcilutite to sequences of calcarenite → "inside shale" → calcarenite alternations and calcilutite → calcarenite → "inside shale". The sequences are the same as those apparent in the system at six iterations, $\alpha \rightarrow \Psi \rightarrow \alpha \rightarrow$

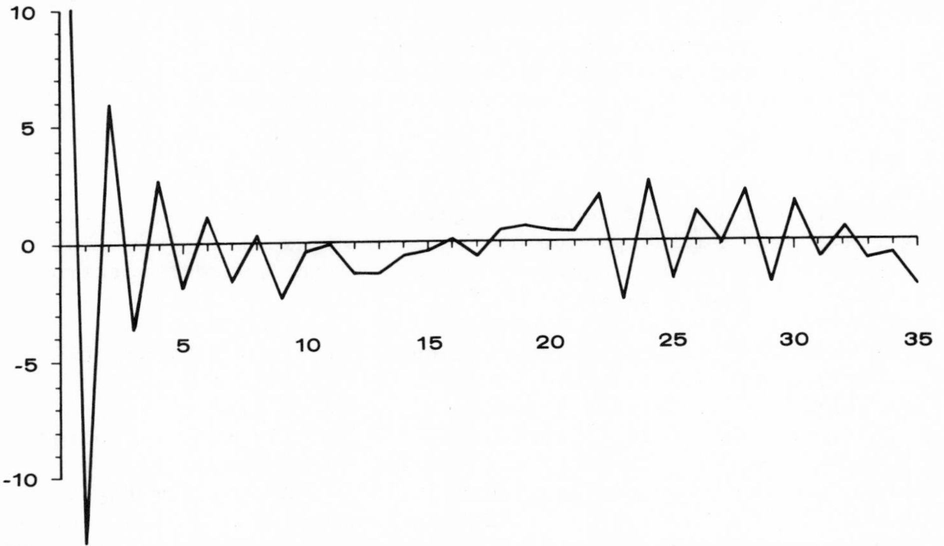

Figure 11. - Autoassociation plot of stratigraphic section after six iterations of combining states by right substitutability.

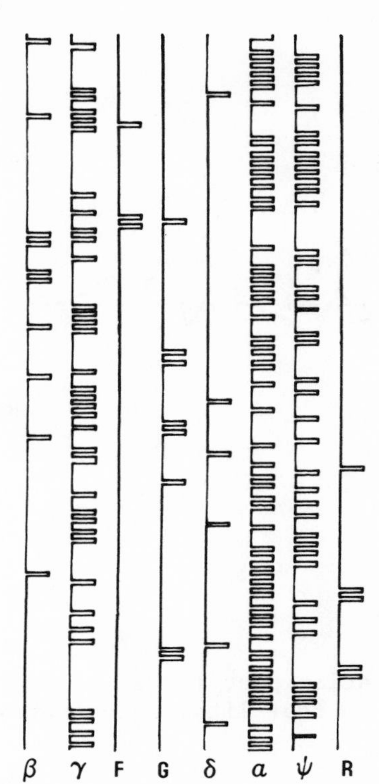

Figure 12. - Diagrammatic section showing occurrences of lithologies after six iterations of combination of states by right substitutability. See Figure 10 for key to states.

Ψ... and γ → α → Ψ → γ → α → Ψ... Although the order γ → α → Ψ ... is dominant, these three states also occur in other combinations. "Outside shales" sporadically appear within this system. Figure 15 shows the succession of four lithic states within the stratigraphic interval.

As a comparison, two runs were made in which lithic types were

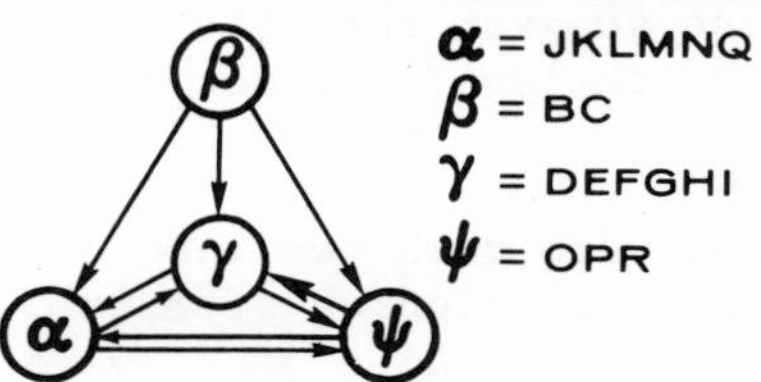

Figure 13. - Flow graph for stratigraphic section after twelve iterations of combining states by right substitutability. Diagram shows pronounced flow involving inside shale, calcilutite, and calcarenite.

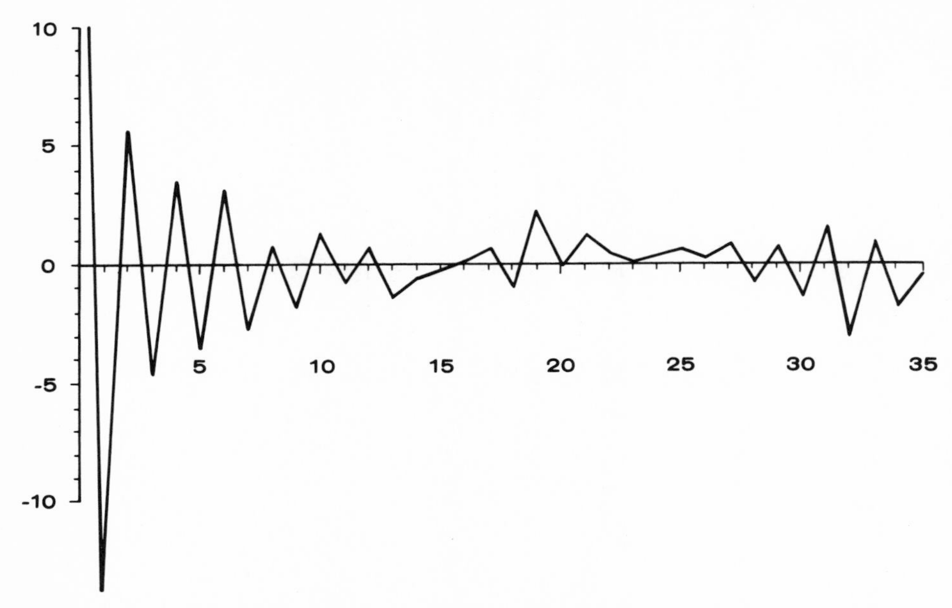

Figure 14. - Autoassociation plot for stratigraphic section after reduction to four lithologic states by right substitutability.

combined into states that correspond closely to those in the classic cyclothem model, and to lithologies used in other mathematical studies of cyclicity. The first of these uses a five state system of an "inside shale", an "outside shale", a black shale, a calcarenite, and a calcilutite. It differs from the model found by left substitutability only in the retention of black shale as a separate lithic category, and in the substitution of pelletal calcarenite for skeletal calcilutite in the calcarenite lithology. The flow diagram (Fig. 16) is similar to Figure 10. Note that the black shale lithology does not contribute to the cycle, but occurs without apparent pattern in the succession (Fig. 17). The plot of autoassociation resembles Figure 14, except for slight shifts in peak positions. These suggest that the distinction of black shales from other "inside shales" does not enhance the appearance of cyclicity.

Next, all limestones were combined into a single lithic category, a procedure used in many studies of idealized stratigraphic successions. "Inside" and "outside shales" were retained as distinct categories, as was black shale. The section oscillates between limestone and "inside shale", with minimal contributions by the other shale lithologies (Fig. 18). The autoassociation dia-

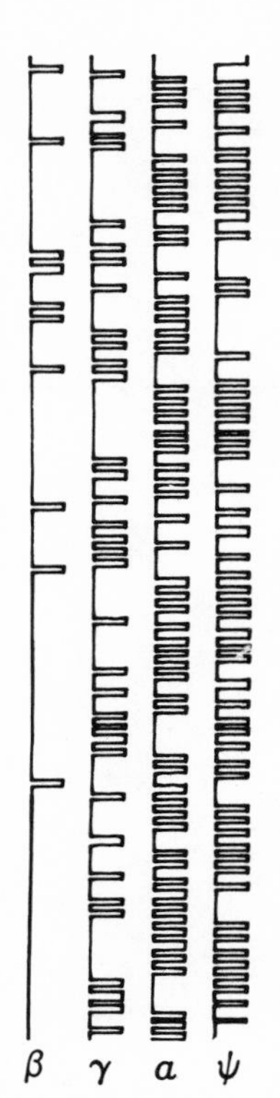

Figure 15. - Diagrammatic section showing occurrence of lithologies after reduction to four states by right substitutability. See Figure 13 for key to states.

gram becomes strongly oscillatory, with a beat frequency of about 16 lags caused by appearance of occasional black shales and "outside shale" (Fig. 19).

The extreme fluctuations apparent in Figure 19 suggest that

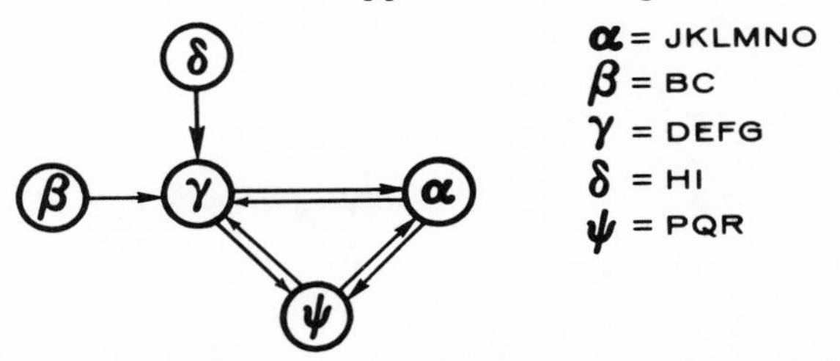

Figure 16. - Flow graph for stratigraphic section with lithologies classified into five arbitrary groups. These closely correspond to Figure 13, except that black shale (δ = HI) is retained as separate category.

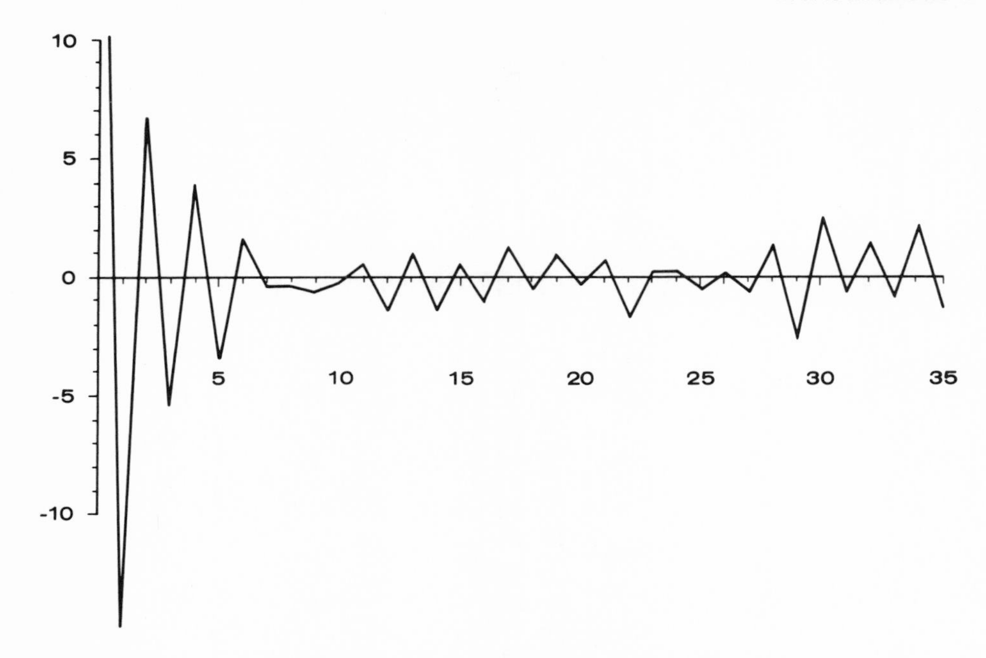

Figure 17. - Autoassociation plot for stratigraphic section containing five lithologic states. Note similarity to Figure 14.

the succession has deteriorated into a disturbed two-state system. Although such a system will of necessity "cycle" as it oscillates between the two states, it is trivial as a model for cyclothems or megacyclothems. The flow diagram and autoassociation plot suggest that the only distinctive roles played by "outside" or black shales are as occasional perturbations to an otherwise stable, alternating system of limestone and marine shale.

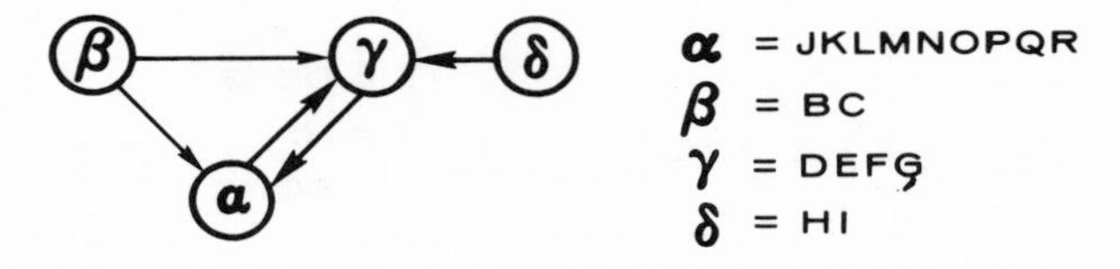

Figure 18. - Flow graph for stratigraphic section after combination of all lithologies into four states: inside shale, outside shale, black shale, and limestone. Oscillation between limestone and inside shale is only pronounced flow pattern.

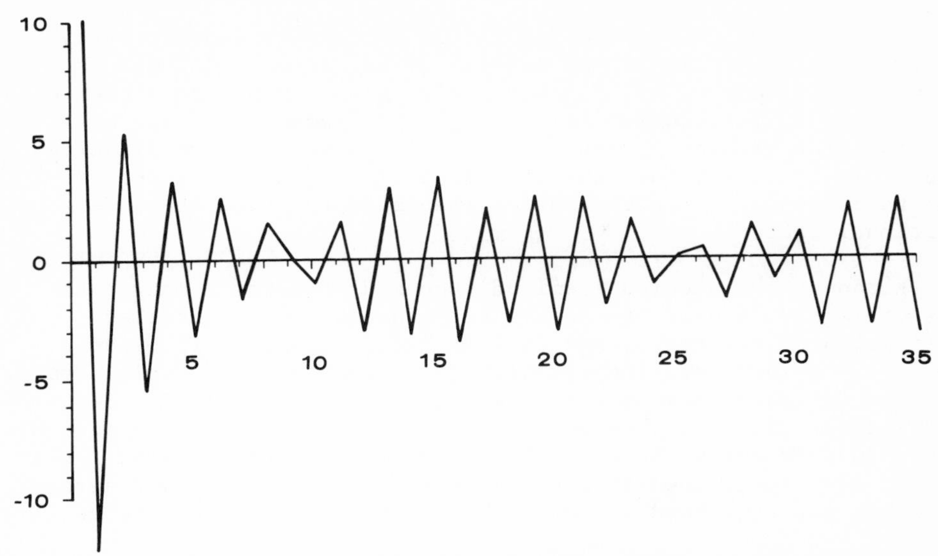

Figure 19. - Autoassociation plot for stratigraphic section after combination of lithologies into four states. Note pronounced beat frequency resulting from perturbation of two-state oscillation between inside shale and limestone.

CONCLUSIONS

The flow graph developed from seventeen-state data for the Missourian section shows no evidence of large cyclic patterns, nor does the plot of autoassociation. Mutual substitutability analysis is capable of reducing the number of iithic states to nine by combining all limestones except algal calcilutite into a single category. However, the shales remain separate, reflecting in part their low probabilities of occurrence and their sporadic pattern of succession. Right substitutability, which is based on upward transition probabilities, provides a more effective criterion for reducing the number of states. After twelve iterations, four lithic categories remain: a calcilutite, a calcarenite, an "inside shale", and an "outside shale". Using these states, the succession reveals a strong tendency for three-state cycles. "Outside shales" occur as sporadic interruptions in this cyclic pattern. The set of lithic states obtained by right substitutability seems at least as adequate as empirical models in which black shales are retained as a separate category.

At the four-state level, the stratigraphic interval measured consists of an alternating series of calcilutites, calcarenites, and marine shales. Sporadic incursions of coarser sediments may interrupt this pattern at any point. We may infer from this that the various limestone and "inside shale" lithologies are intimately interrelated. The occurrence of a specific lithologic type at a certain point within the sequence seems to have no great significance. In contrast, the coarser "outside" shales and siltstones seem to represent a clastic influx independent of the basic depositional pattern.

Although the section examined supposedly contains many examples of megacyclothems, these are not expressed in a detailed examination of the rocks, nor do they become apparent if lithologies are combined into increasingly gross categories. Cyclothems and megacyclothems may be characteristics apparent only in idealized successions, and they may not occur in the actual stratigraphic intervals from which these idealized successions are derived. Alternatively, the recognition of cyclothems may depend upon characteristics not expressed in the lithology of the rocks themselves.

ACKNOWLEDGMENTS

Several persons contributed to the lively debate which accompanied this investigation. These include D.F. Merriam, W. Schwarzacher, P.H. Heckel, and C.D. Conley. Special thanks are extended to Schwarzacher and Conley who re-examined the stratigraphic interval to aid in the confirmation of lithologic assignments. R.C. Moore graciously discussed his concepts of cyclic sedimentation in sessions with members of the Geologic Research Section of the Kansas Geological Survey. A. Rosenfeld provided unpublished documents containing algorithms from which the substitutability measures were programmed. R.J. Sampson wrote the FORTRAN programs and graphic display routines used in the study.

REFERENCES

Anderson, R. Y., 1967, Sedimentary laminations in time-series study, in Colloquium on time-series analysis: Kansas Geol. Survey Computer Contr. 18, p. 68-72.

Anderson, R. Y., and Koopmans, L. H., 1963, Harmonic analysis of varve time series: Jour. Geophysical Res., v. 68, no. 3, p. 877-893.

Anderson, T. W., and Goodman, L. A., 1957, Statistical inference about Markov chains: Am. Math. Stat., v. 28, p. 89-110.

Berge, C., and Ghouli-Houri, A., 1965, Programming, games and transportation networks: Methuen and Co., Ltd., London, 260 p.

Carr, D. D., and others, 1966, Stratigraphic sections, bedding sequences, and random processes: Science, v. 154, no. 3753, p. 1162-1164.

Carss, B. W., 1967, In search of geological cycles using a technique from communications theory, in Colloquium on time-series analysis: Kansas Geol. Survey Computer Contr. 18, p. 51-56.

Duff, P. M. D., Hallam, A., and Walton, E. K., 1967, Cyclic sedimentation: Elsevier Publ. Co., Amsterdam, 280 p.

Krumbein, W. C., 1967, FORTRAN IV computer programs for Markov chain experiments in geology: Kansas Geol. Survey Computer Contr. 13, 38 p.

Krumbein, W. C., and Dacey, M. F., 1969, Markov chains and embedded Markov chains in geology: Jour. Intern. Assoc. Math. Geol., v. 1, no. 1, p. 79-96.

Mann, C. J., 1967, Spectral-density analysis of stratigraphic data, in Colloquium on time-series analysis: Kansas Geol. Survey Computer Contr. 18, p. 41-45.

Merriam, D. F., ed., 1964, Symposium on cyclic sedimentation: Kansas Geol. Survey Bull. 169, 636 p.

Moore, R. C., 1936, Stratigraphic classification of the Pennsylvanian rocks of Kansas: Kansas Geol. Survey Bull. 22, 256 p.

Moore, R. C., 1949, Division of the Pennsylvanian System in Kansas: Kansas Geol. Survey Bull. 83, 203 p.

Moore, R. C., and Merriam, D. F., 1965, Upper Pennsylvanian cyclothems in the Kansas River Valley: Field Conf. Guidebook, Kansas Geol. Survey, 22 p.

Pearn, W. C., 1964, Finding the ideal cyclothem, in Symposium on cyclic sedimentation: Kansas Geol. Survey Bull. 169, p. 399-413.

Preston, F. W., and Henderson, J. H., 1964, Fourier series characterization of cyclic sediments for stratigraphic correlation, in Symposium on cyclic sedimentation: Kansas Geol. Survey Bull. 169, p. 415-425.

Rosenfeld, A., Huang, H. K., and Schneider, V. H., 1968, An application of cluster detection to text and picture processing: Univ. Maryland Computer Science Center, College Park, Md., Office Naval Res. Grant Nonr 5144(00), Tech. Rept. 68-68, 64 p.

Sackin, M. J., and Merriam, D. F., 1969, Autoassociation, a new geological tool: Jour. Intern. Assoc. Math. Geol., v. 1, no. 1, p. 7-16.

Schwarzacher, W., 1964, An application of statistical time-series analysis of a limestone-shale sequence: Jour. Geology, v. 72, no. 2, p. 195-213.

Schwarzacher, W., 1967, Some experiments to simulate the Pennsylvanian rock sequence of Kansas, in Colloquium on time-series analysis: Kansas Geol. Survey Computer Contr. 18, p. 5-14.

Schwarzacher, W., 1969, The use of Markov chains in the study of sedimentary cycles: Jour. Intern. Assoc. Math. Geol., v. 1, no. 1, p. 17-39.

Sokal, R. R., and Sneath, P. H. A., 1963, Principles of numerical taxonomy: W. H. Freeman and Co., San Francisco, 353 p.

Vistelius, A. B., and Feigel'son, T., 1965, The theory of formation of sedimentary beds: Doklady Akad. Nauk SSSR, v. 164, no. 1, p. 158-160.

Weller, J. M., 1930, Cyclical sedimentation of the Pennsylvanian Period and its significance: Jour. Geology, v. 38, p. 97-135.

Weller, J. M., 1960, Stratigraphic principles and practices: Harper & Bros., New York, 725 p.

Weller, J. M., and Wanless, H. R., 1939, Correlation of minable coals of Illinois, Indiana, and western Kentucky: Am. Assoc. Petroleum Geologists Bull., v. 23, p. 1374-1392.

Zeller, D. E., ed., 1968, The stratigraphic succession in Kansas: Kansas Geol. Survey Bull. 189, 81 p.

MATHEMATICAL MODELS FOR HYDROLOGIC PROCESSES

G. de Marsily

Ecole des Mines de Paris

ABSTRACT

Models of simulation of groundwater flow in sedimentary basins are generally used for water-resources optimization problems; however, a byproduct of their building may be additional information on the structure of the basins themselves.

In the first stage of the simulation procedure, the hydraulic parameters of the sediments (permeability, porosity) have to be determined, using lithostratigraphical and hydrological data. The values of these parameters are closely related to the sedimentation process itself.

Two different techniques of identification of these parameters are presented. Practical examples on Tertiary marine sediments in Aquitaine, and Quaternary alluvial sediments, will show the complement of information gained in this manner on the structure of the basins.

INTRODUCTION

Models of simulation of the behavior of water in a system have been developed for several years, and are, to some extent, known to all geologists (Emsellem, 1970; de Marsily, 1972 ; Matalas, 1969; Prickett and Lunnquist, 1968 ; Remson and others, 1971; Weber, Peters, and Frankel, 1968; Witherspoon, Javandel, Neuman, 1968). Their general purposes (Kisiel, 1971) are to predict the future state of the system several years hence, under various conditions of exploitation (controlled input), while assuming the most likely values for the uncontrolled input (rainfall, recharge, etc...). One may even want to optimize the exploitation of the system according to a given objective function (maximum benefit, etc.). This may be obtained by performing different simulations, with different conditions of exploitation, or different equally probable uncontrolled inputs. Optimization techniques also may be used. Finally, they help to decide how systems can be used, or which improvements are to be added (reservoirs, artificial recharge, waste-water reclamation, etc.).

This paper is restricted to the simulation of the flow of groundwater into aquifers, and the state of the system will be described only by the hydraulic head of groundwater. After a brief presentation of the methods of simulation in current usage, we will turn to the most important problem of fitting of such models, i.e. the determination of the set of parameters that will make the model behave like the actual system. Two of these parameters, permeability and storage coefficient, are closely related to the sedimentation process of the strata where the flow occurs. Evidence of the interaction of hydrology and sedimentology in this area will be discussed.

DIGITAL MODELS OF WATER FLOW IN AQUIFERS - PRINCIPLES

This simulation technique is based upon two physical laws concerning the flow of water through porous media.

Darcy's Law

Considering a monodimensional flow, and an element of porous medium, the direction and the velocity of the flow can be determined when the hydraulic heads on each side of the element are known,

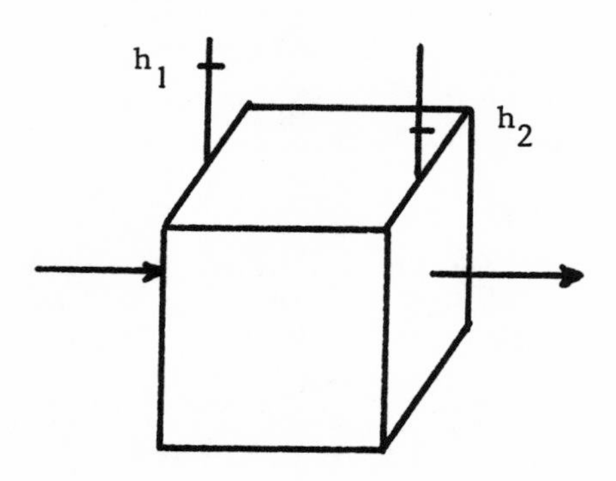

$$\vec{V} = -K \, \overrightarrow{grad} \, h = K(h_1 - h_2) ,$$

where $\vec{V}$ = the velocity vector,

K = the permeability in the direction of the flow, and

h, h_1, h_2 = hydraulic heads.

This relation is valid only if the entire element is saturated with water.

Continuity Equation

It is the simple application of the general mass conservation relation. Considering a multidimensional flow, and an element of porous medium where this flow occurs, it states that the sum of the flows entering through each side of the element is either null (steady state) or equal, for a given period Δt, to the quantity of water stored or released by the element of porous medium.

This ability of the element to store or release water is obtained by an increase or decrease of its hydraulic head following two possible mechanisms.

(1) If the aquifer is confined by an impervious layer, and if the porous medium is saturated with water, then the combined compressibility of the water and of the porous medium itself permits this storage of water. A storage coefficient S is defined (nondimensional), equal to the quantity of water that can be stored or released by a unit change of hydraulic head in a volume of porous medium having the total thickness of the aquifer as height, and a unit surface.

(2) If the aquifer is unconfined, then the level of water in the porous medium (in other words, the hydraulic head) will rise or fall, and water will be stored or released according to the porosity of the porous medium. In this situation, the storage coefficient also is used, but is equal to the porosity.

Those two laws make it possible to write in mathematical form the equation governing the flow of water in an aquifer.

Analytically, it gives rise to the well-known diffusivity equation. Considering a one-layer aquifer, this is written

$$\text{div} \, (T \, \text{grad} \, h) = S \frac{\partial h}{\partial t} + q, \text{ or} \qquad (1)$$

$$\frac{\partial}{\partial x} T_x \frac{\partial h}{\partial x} + \frac{\partial}{\partial y} T_y \frac{\partial h}{\partial y} = S \frac{\partial h}{\partial t} + q,$$

where T = the transmissivity of the aquifer (thickness or saturated heights of the aquifer multiplied by its permeability K), defined in each point (x,y) of the plan;

T_x, T_y = transmissivities in the direction x or y of the plan, if T is not isotropic;

h = hydraulic head of the water, defined in each point x,y;

S = storage coefficient, defined in each point x, y; and

q = flow rate withdrawn or injected by unit surface of the plan between the aquifer and the outside medium: pumpage, for instance.

Speaking in terms of digital simulation of aquifers, it is unnecessary to carry out any part of this analytical formulation. A one-layer aquifer is divided by a grid into a certain number of elements, and the massconservation law plus Darcy's law yield a set of linear equations relating the head in each element of the aquifer to the transmissivities, storage coefficient and withdrawn flowrate (pumpage, etc...) in each element, as well as to the boundary conditions imposed on the hydraulic head at the limits of the aquifer.

Using, for instance, a 5-point finite difference explicit formulation on a square grid, the equation for an element C writes as follows

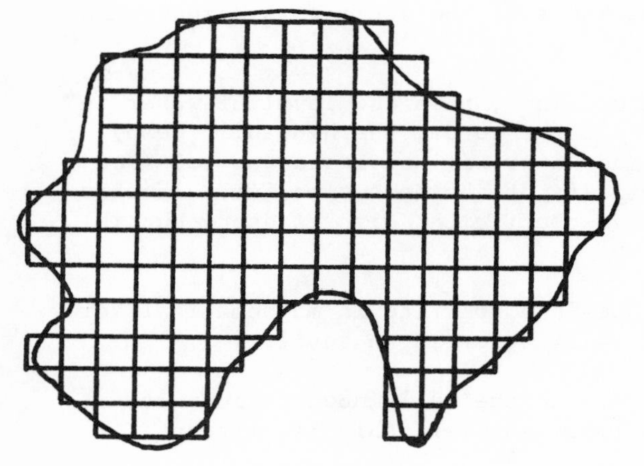

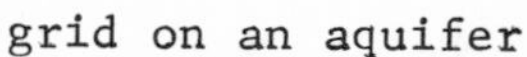

grid on an aquifer

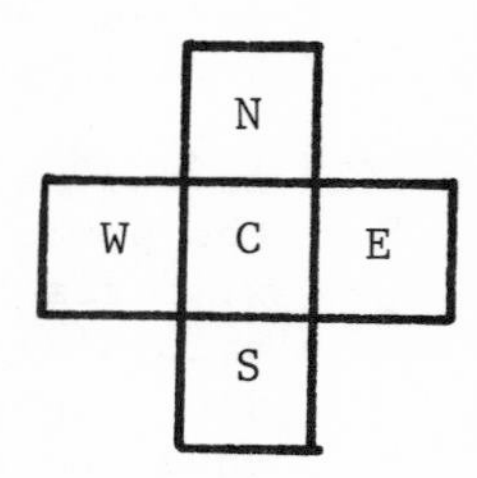

five adjacent elements of the grid

$$\frac{T_N+T_C}{2}(H_N-H_C) + \frac{T_S+T_C}{2}(H_S-H_C) + \frac{T_E+T_C}{2}(H_E-H_C) + \frac{T_W+T_C}{2}(H_W-H_C) =$$
$$S_C a^2 \frac{H_C^*-H_C}{\Delta t} + Q_C, \qquad (2)$$

where T_C, T_N, T_S, T_E, T_W = transmissivities in each element C, N, S, E, W;

S_C = storage coefficient in element C;

H_C, H_N, H_S, H_E, H_W = hydraulic heads in each element at time t;

H_C^* = hydraulic head in element C at time t + Δt;

Δt = time increment;

a^2 = area of each element of the grid; and

Q_C = sum of the flow rate exchanged in element C with the outside.

Such an equation can be written successively in each element of the grid.

This gives a set of equations that can be written in matrix form for the entire aquifer

$$\bar{\bar{T}}H = \bar{\bar{S}}\,\frac{H^*-H}{\Delta t} + Q, \qquad (3)$$

where $\bar{\bar{T}}$ = a matrix whose coefficients are a linear function of all the transmissivities in each element of the aquifer;

$\bar{\bar{S}}$ = a diagonal matrix whose coefficients are the storage coefficient in each element of the aquifer;

H, H^* = vectors whose components are the heads in each element, at time t and t + Δt; and

Q = a vector of the flow rates exchanged in each element with the outside.

It must be pointed out here that the approximation given by this model to the reality mainly depends on the size of the grid. Heads, which are supposed to be known at time t and unknown for time t + Δt, represent average values of the actual heads taken over each element of the grid. Accordingly, transmissivities and storage coefficients are not supposed to be constant, and may be allotted different values in each element, as far as those values can be determined. Each value will represent an average over the element of the punctual value of the parameter. In this discrete manner, heterogeneity and anisotropy (supposed to have principal directions along the sides of the grid) can be taken into account.

Boundary conditions given on the heads (constant potential = fixed head), or known flow rates Q at a limit, are introduced into the equations such as (2) for the elements of the grid forming the limit of the aquifer.

Running a simulation will start from an initially known distribution of heads, by solving system (3), thus determining heads at time $t + \Delta t$, which are considered as data for solving again system (3) in order to determine heads at time $t + 2\Delta t$,... and so on.

In the explicit method, there is a constraint on the size of the time increment Δt, which generally makes it necessary to perform this sequence a great number of times to determine the distribution of heads in the future.

If the steady state of the system is desired, i.e. the head distribution when the system has reached an equilibrium and ceases to vary with time, then system (3) evolves into

$$\bar{\bar{T}}H = Q, \qquad (4)$$

which also can be solved to compute the final head distribution H. In that situation, neither storage coefficient nor initial heads need to be known.

There are other formulations as well (implicit, Crank-Nicholson, etc.) and even other discretizations (finite elements, etc.) by which the same results may be obtained through different numerical techniques.

A last remark should be made concerning the representation of multilayered aquifers, which are commonly encountered in large sedimentary basins. Two categories of layers are distinguished:

(a) pervious layers, in which the flow will be considered mainly horizontal, or parallel to the stratification, and which will be represented as the two-dimensional aquifers studied;

(b) semipervious layers, in which the permeability is too low to allow significant horizontal flow, but through which vertical flow can occur (or rather a flow orthogonal to the stratification) in a significant manner because of the large contact area compared to the thickness of the layers. These layers will not be represented in the models as aquifers, but as links between aquifers, or capacities overlaying aquifers, and able to transmit or produce water.

The real system being a superposition of such layers, the model will consist of a superposition of two-dimensional aquifers as described by equation (2), representing the pervious layers, but connected with each other directly, or through semipervious layers, by the top and bottom sides of each element of the superposed grids. That means that in the left side of equation (3), additional terms representing the vertical flow from one layer to the adjacent ones are taken into account.

This yields a new system similar to (3). This technique is most commonly used. There are however, three-dimensional representations, which may also include the unsaturated flow (Freeze, 1971).

FITTING SIMULATION MODELS OF AQUIFERS

To achieve this, that is to say to make the model behave as much as possible like an actual aquifer, different types of data are needed.

Hydrological Data

A known behavior of the aquifer must be recorded with all the hydrological input needed (pumping rate, alimentation through rain or recharge, estimation of evapotranspiration, etc.)

If a steady state of the system can be observed, the fitting is made easier, because only the transmissivities need to be determined. A piezometric chart of this steady state must be drawn, in this situation, whereas in transient state, the data covering several years of behavior may have to be available.

Boundary conditions also must be known - or guessed - at the limit of the aquifer: either constant potential, which means superabundant recharge of the aquifer, or no flow, which means contact with an impervious layer (fault or vanishing of the layer), or prescribed flow, which can be contact with another aquifer whose yield will not vary too much with time, whatever the conditions of exploitation.

Finally, some punctual values of the transmissivities, storage coefficients and permeabilities of semipervious layers can be known through hydraulic determination by performing local pumping tests in the wells, or by interpreting natural variations of the system throughout the year through analytical solutions (relations between the level in a river and the head of an aquifer may yield estimates of the diffusivity T/S, for instance)(Baetz and others, 1967; Emsellem and others, 1971).

But, the values of these parameters are generally not in sufficient number to estimate their distribution in the entire system.

Geological Data

They are even more important than hydrological data, and much more difficult to obtain. They may be classified under two labels:

(1) Structure. - Refers to the geometry, extension and relative positions of all the strata of the basin where water may flow. These may be porous (sandstone) or permeable by fractures, e.g. limestone, or semipervious (clay, marl). It must be pointed out again that layers generally considered by geologists as impervious, must sometimes be taken into account in the simulation of large systems, because even if their permeability is low, nonnegligible flow may occur through them if they are not too thick and are in contact over large areas with aquifers (for instance, over a basin of 10,000 km^2, a 10 m layer of clay of permeability 10^{-10} m/sec, separating two aquifers having a difference of head averaging 10 m over the basin, may leak as much as 1 m^3/sec, that is to say enough water to satisfy the needs of a city of 150,000).

(2) Substance - Consists of adding to the structure estimated values of the permeability and storage coefficient of each layer. This can be done by giving a single value, or a minimum and a maximum admissible value, or even a frequency distribution of the likelihood of a set of values.

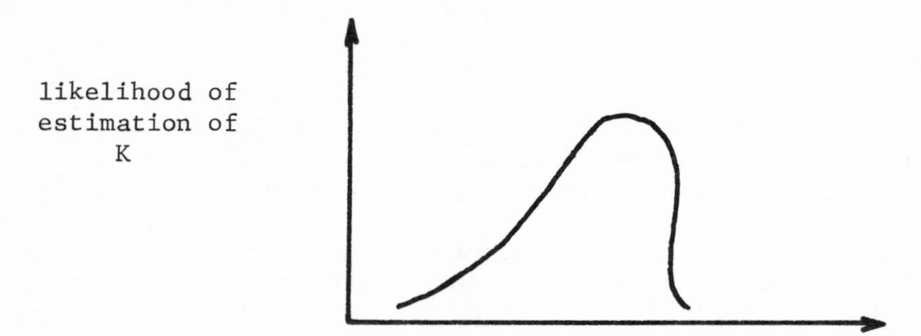

To achieve this, all the branches of geology may have to contribute:

- stratigraphy and paleontology for correlations between wells, which may use trend-surface analysis (Harbaugh and Merriam, 1968; Matheron, 1969; Matheron and Huijbrechts, 1970).

- lithology for estimating parameters through sample description,

- geophysics for both structure and substance,

- well logging, for more knowledge on the data points,

- sedimentology, to build better extrapolation from some punctual values, according to the history of sedimentation, the gradients of sediment thickness and of their granulometry, main features such as location of rivers, delta, etc., this may also involve the use of geomorphology.

Clearly, an information system would be needed to build a coherent synthesis of all geological data.However, all unknown parameters cannot be estimated in this manner. Thus, the simulation model will have to fulfill this need by combining hydrological and geological data. This can be done either by traditional trial and error methods or automatic fitting, known as the inverse problem.

TRIAL AND ERROR METHODS

Having assigned values to all known and guessed parameters, a simulation is performed, and the calculated heads are compared to the actual ones. Major differences generally occur, and to balance them, modifications are made on the parameters, if possible within the prescribed limits of variation, and another simulation is run. Better agreement then should be obtained (if not, the wrong parameters were modified, in which situation the geologist may have to give his notice!). Repeating this procedure, agreement can generally be obtained, sometimes at the cost of major modifications of the model: boundary conditions, connections between aquifers, values of parameters, etc. Additional data may be needed to check with possible alternatives.

To illustrate this, an example will be shown from the digital model of the Tertiary aquifers of northwestern Aquitaine, simulated in the Ecole des Mines at Fontainebleau.

The Aquitaine model was the result of an important joint work undertaken by several administrations, laboratories and companies under the scientific leadership of Prof. H. Schoeller (Schoeller and others, 1969, 1971). Briefly the aim of this model was to estimate the consequences of an increase of the pumping rate in the aquifers near the city of Bordeaux, needed for the city development. As Prof. Schoeller initially guessed, the model proved the existence of a limit in this withdrawal rate, already nearly reached, and beyond which there were serious risks of contamination of the aquifers by salt water, mainly from the Gironde (end of the Garonne River, nearly as salty as the sea).

This could turn into national disaster, because this region contains the world famous vineyards of Medoc, where the intrusion of salt water would mean irrepatable destruction (a list of the first classified vineyards of Medoc is given in the Appendix).

The Tertiary sediments of Aquitaine lie on the Cretaceous limestone, considered an approximately impervious substratum. From the Eocene, through Oligocene, Miocene to the Plio-Quaternary formations, it consists of a complicated succession of gravels, sands, limestones, marls and clays with all intermediate facies, some limited in extension, some present over large areas. It reflects the numerous transgressions and regressions of the sea during the Tertiary. No continuous impervious layer can be found, and the entire series must be considered as an aquifer system, sometimes as much as 500 m thick. The area studied covers about 15,000 km^2; 16 cross sections were made, at a grid size of about 20 km, with lithostratigraphic interpretation (Marionnaud, 1967, pl. 1, 2).

Drastic simplifications were necessary to cut the number of layers to 5 main aquifers, linked together either horizontally, vertically or through semipervious beds. (The last model contained over 700 nodes.) To shorten this presentation already published elsewhere (Schoeller and others, 1971), we will concentrate on the main aquifer, the sands of the lower Eocene S2.

The fitting was performed by trial and error, on a steady state measured in 1965 due to regulation of the pumping rate since 1959 (Fig. 1). The initially guessed distribution of transmissivities of S2 (Fig. 2) was based on all existing data, and previous hydraulic studies of this aquifer. It showed a somewhat circular distribution of the transmissivities, in the southeast of the aquifer, with high values towards the center, and lower values towards the limits. The southern limit of the aquifer is in fact arbitrary, for it has further extension southward, but was imposed hydrogeologically to limit the extension of the model. The northern limit is outcrop.

Rather important modifications of this distribution, together with other modifications in the different aquifers, were necessary to obtain good agreement between actual and computed heads.

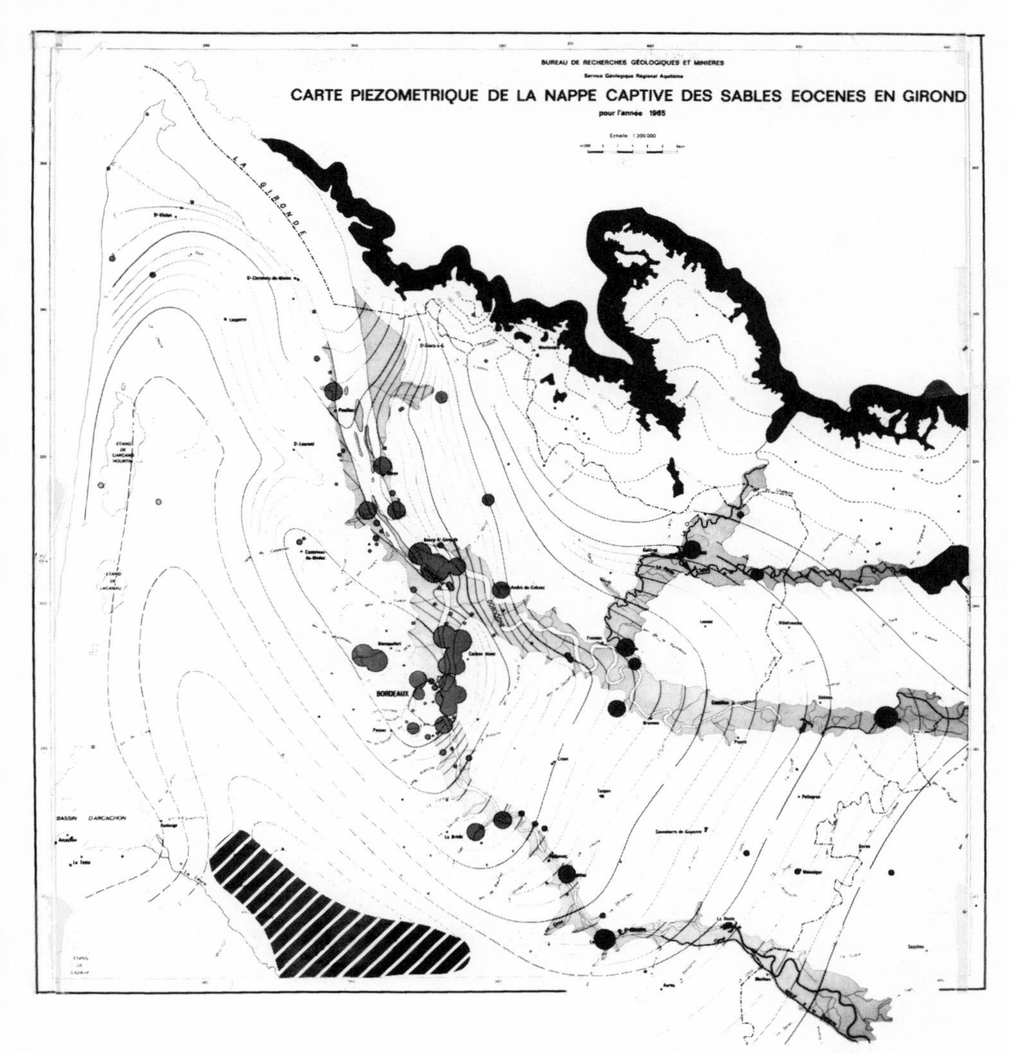

Figure 1. - Piezometric chart of aquifer of lower Eocene sands, S2, in Gironde. Isopiezometric lines contoured every meter of head; circles indicate localization and importance of pumping (BRGM, 1965).

Figure 3 shows the new distribution of transmissivities of S2, obtained after the fitting. The central area of high transmissivity is now more or less shaped like a Y, with one branch going in the same direction as the river Garonne, and the other going more east. High transmissivity being associated with coarse sands or gravels,

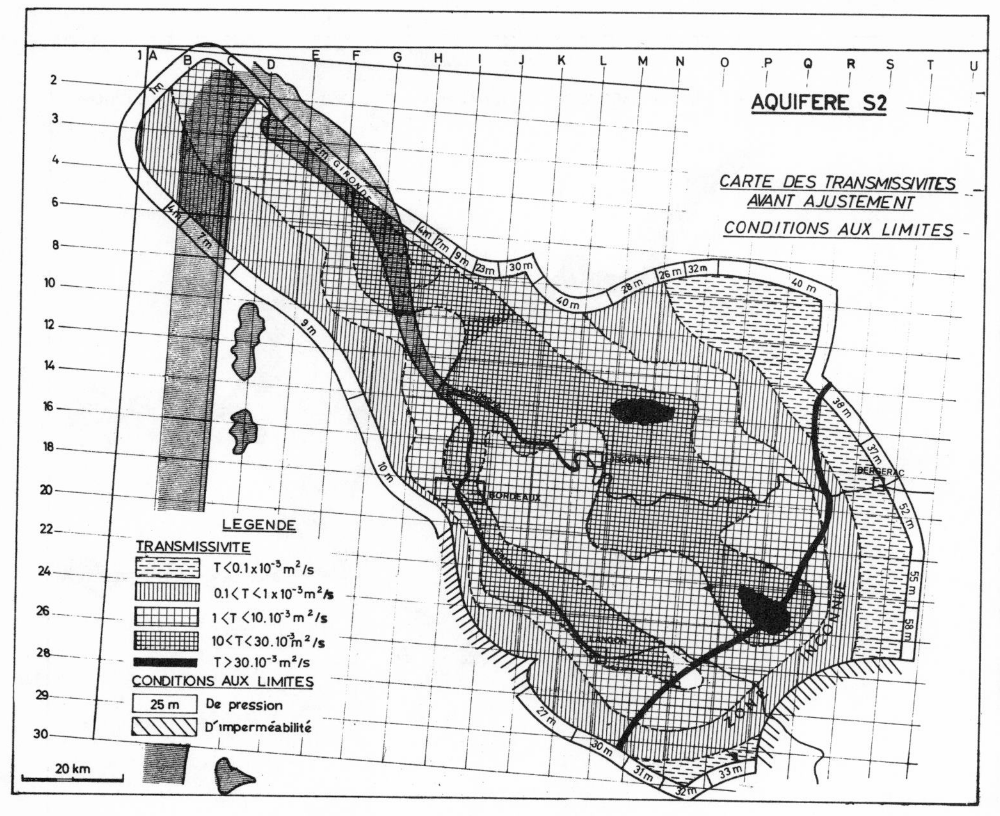

Figure 2. - Distribution of transmissivities as initially estimated in aquifer of lower Eocene sands in northern Aquitaine.

or greater thickness of the layer, this gives a different image of the sedimentation process, which may have received materials from two leading outlets. This information was confirmed by geological reinterpretation of the data, the latter transmissivity map being more rational than the former.

To finish with Aquitaine, it may be added that a transient fitting also was performed, with 100 years simulation from 1870 (around the initial drilling to the artesian aquifer S2), with good agreement with the existing data. It was shown that the transient period of the system was short (about 2 years) due to the considerable leakage of the aquitards. The final decision was to limit the increase of the withdrawl in the area of Bordeaux, so that you may hope to drink Medoc wine for a while!

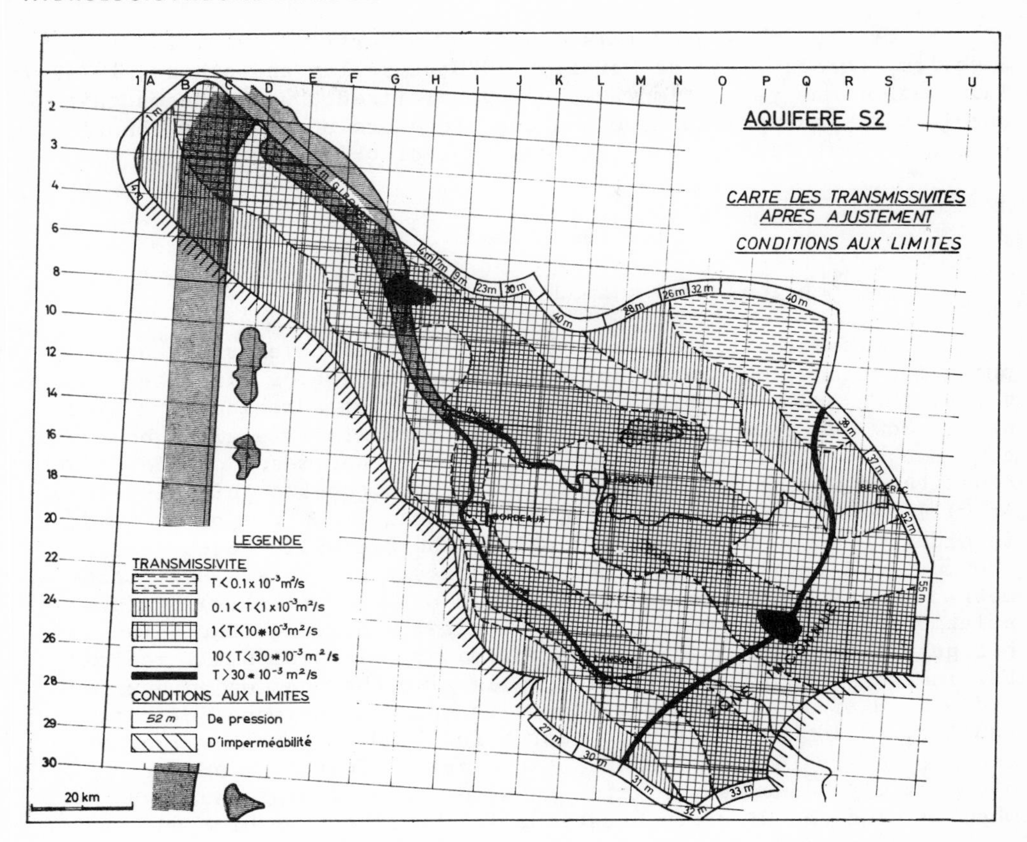

Figure 3. - Distribution of transmissivities after calibration of model in aquifer of lower Eocene sands in northern Aquitaine.

AUTOMATIC FITTING

In order to avoid this trial and error fitting, that can prove to be difficult, lengthy and expensive, several authors have tried to find a method of determining parameters automatically. This is known as the inverse problem; the direct problem being to simulate the heads when the values of the parameters are known, the inverse problem is to compute the parameters when the heads are known (Coats, Dempsey, and Henderson, 1970; Jacquard and Jain, 1965; Kleinecke, 1971; Nelson, 1970).

Mathematically speaking, the problem is said to be improperly posed, and is rather difficult to solve. As an example, the new inverse method developed at the Ecole des Mines in Fontainebleau will be briefly presented (Emsellem and de Marsily, 1969, 1971; Emsellem, Lerolle, and de Marsily, 1970; Emsellem and others, 1971). This method may yet be improved and generalized, but it seems that the initial assumptions made and the procedure adopted may constitute a good starting point for this type of problem.

Inverse method

Two assumptions must be made:

(1) transmissivities and even storage coefficients are not point-functions, but set-functions (Matheron, 1967). In other words, they are not defined punctually but over a region of a given size. Now, when two values of the same parameter are given on two continuous regions, the average value of the parameter on the block constituted by the two regions is related to the two initial values. It will be approximated here by their average. On the other hand, in order to solve the inverse problem, the hydraulic heads needed are measured in a certain number of observation wells. The piezometric chart then is contoured (by hand or automatically) by interpolation between the wells, and geological knowledge. But this does not permit the estimation of the parameter values of small regions. For instance, if 100 observation wells give the head on an aquifer, and a grid of 1000 network elements is used on the model of this aquifer, it is unrealistic to search for 1000 different values of the parameters in each element of the grid. (This high number of elements may be needed to get a good approximation of the heads when solving the discrete equations in the direct problem and is not a priori related to the number of observation wells.) The estimation obtained would be highly hypothetical, and more related to manipulations than to real data, 100 or 200 different values would be more realistic.

Finally, it has been observed that more usual methods of solving the inverse problem give rise to oscillating values of the parameters between elements or regions. This is due to the presence of a kernel in the mathematical formulation of the problem, and should be avoided systematically. (The existence of a parasite solution which may be added to the true solution with more or less weight, and makes it oscillating.)

Consequently, the procedure of determination will be by successive approximations, starting from an initially uniform value of the parameters over the entire aquifer, then subdividing the aquifer step by step into smaller regions (defined either by the geology of the strata, or by neutral divisions: 1 region, 4 subregions of equivalent area, 16 subregions, 64, etc.). At each step, the parameters will be optimized in order that:

- the new values allotted to each new subregion will have an average equal to the unique value allotted to the unique region before subdivision in the preceeding step. For instance, if one transmissivity T_1 was computed at the first step, and if at the second step 4 subregions are defined, the 4 different values of transmissivities in the 4 subregions will have T_1 as average;

- the computed distribution of the parameters for a given step will be as close as possible to uniformity, i.e. the solution will be as smooth as possible, and irregularity will be allowed only when absolutely necessary.

(2) The piezometric heads used in the inverse problem, the results of measurements and extrapolations, are likely to contain errors. Furthermore, it is known that any punctual modification of a parameter in a model will have a consequence not only on the vicinity of the given point, but sometimes over the entire aquifer. Consequently, the inverse method will not allow any local modification of the parameters, in order to prevent any punctual error in the given head from creating a locally wrong estimation of the parameters. The modification of the values of the parameters from one step to the next will be done simultaneously in the entire aquifer, and the agreement judged on the entire aquifer as well.

Finally, the quantity to minimize during the optimization procedure will not be the differences between actual and computed heads, but the error flow, i.e. the artificial flow that should be injected or withdrawn in each element of the grid, in order that the model behave exactly as the aquifer, for a given step of the determination. This has four advantages:

- it avoids any direct simulation of the head between two steps of the inverse method, thus minimizing the computer time;

- it limits the influence of the errors in the heads;

- it is more in the line of the final use of the model, when the available resources will have to be determined, in terms of flows;

- it makes it possible to stop the procedure at a given size of subdivision, either in the entire aquifer or in a given region, when these artificial error flows are of the same order of magnitude as the actual admissible uncertainties in the flow entering into each element. These uncertainties may be evaluated for instance, from the uncertainty on alimentation of withdrawl (rain or evaporation), or even from the uncertainty on the measured heads (which can be transformed into terms of flow for a given value of the parameters, which can be done in the inverse method by using the values of the preceeding step).

With these assumptions, an algorithm was built. Consider for the sake of simplification, the situation of a steady-state aquifer of only one layer. The transmissivities will have to be determined. The first step is to compute an optimal uniform transmissivity for the entire aquifer. This single value will be determined by a minimization procedure built as follows. If T is the actual distribution of transmissivities to be used, equation (4) should be satisified

$$\bar{\bar{T}}H = Q ,$$

instead, if T' is another distribution of transmissivities, equation (4) will not be satisfied, generally speaking

$$\bar{\bar{T}}'H \neq Q ,$$

but, one may add to the right side of this equation a new term Q', regarded as an error flow vector, in order that the equation balance,

$$\bar{\bar{T}}'H = Q + Q' . \qquad (5)$$

This error flow vector Q' has a component in each element of the network. These components may be considered as the flow that should be injected or withdrawn in each element to balance the equation. In other words, it represents in term of flow the error made by estimating the actual transmissivity distribution T by the approximation T'.

Now, if T' is the initial homogeneous distribution of transmissivity at the first step, it is possible to isolate Q' in equation (5) and to express its value as a function of T' only

$$Q' = - Q + \bar{\bar{T}}'H .$$

T' then can be computed by cancelling the derivative with respect to T' of the sum of the squares of this error flow Q' in each element of the grid. This is known as least-squares optimization. Practically, another function of T' is also minimized with $\Sigma Q'^2$, in order to insure maximum smoothness of the solution in subsequent steps.

This initial homogeneous transmissivity will not, generally speaking, be sufficient. A division of the aquifer into a small number of regions, say four, will be decided. The new estimation of transmissivities will consist of four different values, with an homogeneous transmissivity in each region. Minimization will take place on the error flow associated with this new distribution on the entire aquifer, and the four transmissivities will be determined simultaneously.

At this step, if necessary, a new division of the aquifer will be decided, say in 16 subregions, and the same procedure will be applied again. From one step to the next, a homogeneous region can be divided into several parts, but the mean values of the estimations in each part must be equal to the preceeding homogeneous estimation. The 16 parameters will be determined simultaneously, and subdivision will again take place until the error flows, which get smaller and smaller with each minimization, have reached an acceptable order of magnitude. A practical example will show the use that can be made of this new technique.

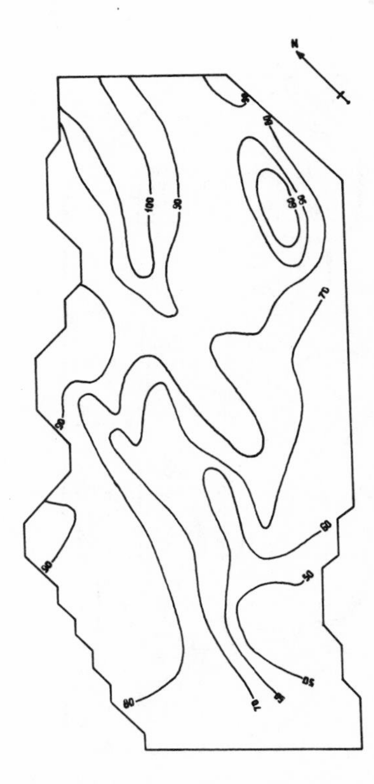

Figure 4. - Isodepth of bottom of aquifer in Rhine ditch, as measured by geoelectric prospection. Depths are given in meters. Right limit is the Rhine, left is outcrop of adjacent Pliocene aquifer, and top and bottom are arbitrary limits in alluvial aquifer. (Document SGAL).

Rhine, Alluvial Aquifer, North of Strasbourg

This study was initially made by the Service de la Carte Geologique d'Alsace et de Lorraine and the Company "Geohydraulique" on an electric analog model. Later, it was transformed into a digital model at Fontainebleau.

The aim of the study was to predict the behavior of this important aquifer under various stresses (new pumpage, pollution hazard, public work on the Rhine for hydropower). The aquifer consists of coarse alluvial sediments (sands and gravels) of relatively high thickness (about 80 m) lying in the Rhine ditch over impervious Oligocene marls. It receives lateral recharge from the adjacent Pliocene aquifer of rather low permeability.

A geoelectric prospect had given the position of the substra-

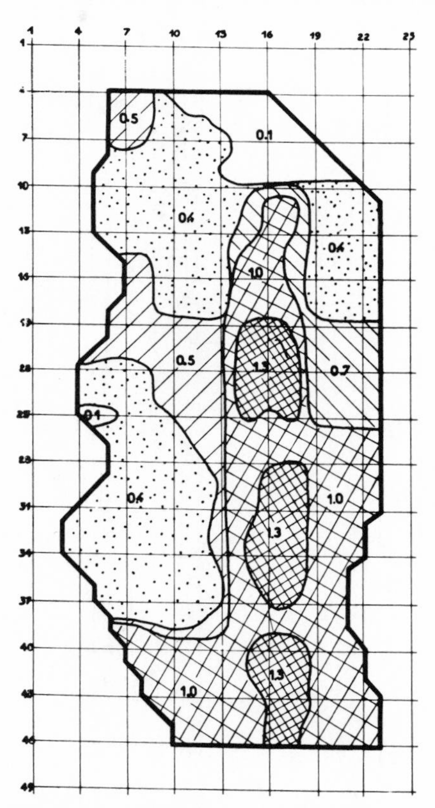

Figure 5. - Transmissivities as determined by inverse method, in 10^{-1} m^2/sec. Numbers on top and left indicate numbers of grid of mathematical model. (Document Ecole des Mines).

tum of the aquifer (Fig. 4) as well as the apparent resistivity of the sediments. The portion of the aquifer studied on the model was limited by the Pliocene terrace on the west side, by the Rhine on the east side (hydraulic limit), and by artificial limits on the north and south sides, far from the main points of interest. Although the heads of the aquifer follow seasonal variations, a quasisteady state can be observed at the end of the summer. Such a state was recorded in October 1966, as well as all possible hydrologic data (flow in the river, in the wells, etc.).

The inverse method was tried on this set of data, with the result shown on Figures 5 (transmissivity map) and 6 (agreement between observed and computed heads in the aquifer). Its main feature is an axis of high transmissivities in the center of the region, probably related to coarser sediments laid in the former

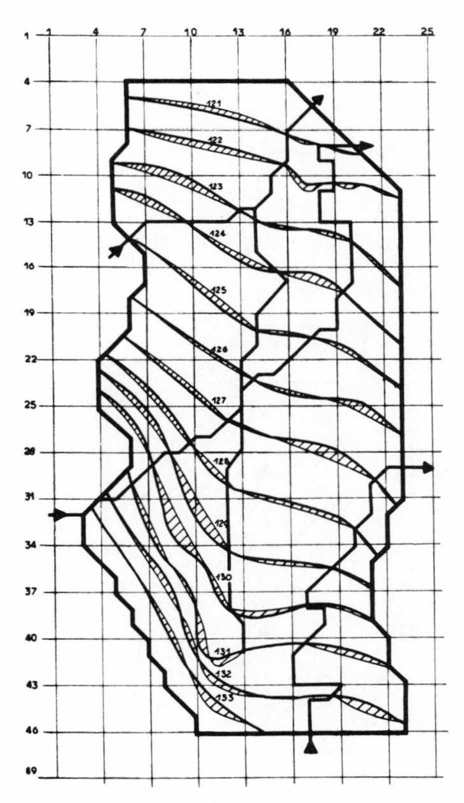

Figure 6. - Agreement between observed and computed heads in aquifer. Isopiezometric lines are in meters, thick lines are observed and thin lines are computed, difference between them is hachured. Small rivers are indicated by arrows. (Document Ecole des Mines).

path of the Rhine. In other contexts, high transmissivity also could be associated with higher tectonization of a fractured rock, but this is not the case of alluvial sediments.

This transmissivity map could be checked by using another method of studying sediments: correlation between electric resistivity, as measured by electrical prospect, and permeability (Duprat, Simler, and Ungemach, 1969; Ungemach, Mostaghimi, and Duprat, 1969). The water of the aquifer was sampled on numerous sites, and its resistivity $\rho\omega$ measured.

When the actual resistivity of the formation ρ_f is measured, Archie's formula shows that

$$\rho_f \; s^2 = \rho\omega \; \phi^{-m},$$

where s is a coefficient of saturation, equal to 1 in the aquifer, ϕ is the porosity and m a constant coefficient known as the cementation factor.

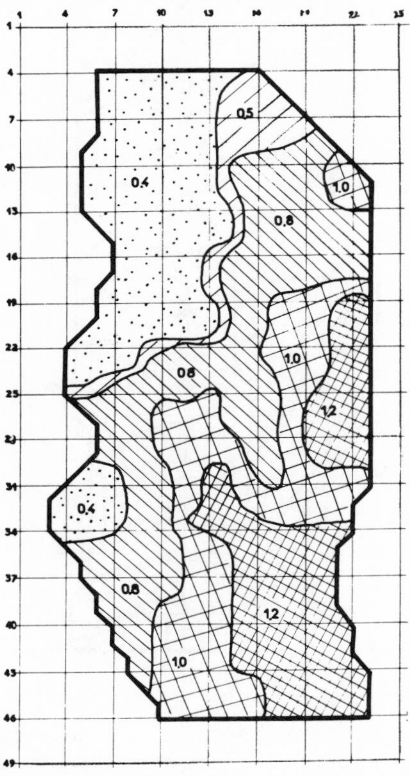

Figure 7. - Transmissivities as determined by correlation with transverse resistivity measured by geoelectric prospect, in 10^{-1} m^2/sec. (Document SGAL-Ecole des Mines).

It is possible then to determine the resistivity of the formation ρ'_f that would be obtained with an aquifer having water of constant resistivity. The corrected transverse resistance of the formation then is computed

$$R'_t = \rho'_f \cdot h,$$

where h is the thickness of the aquifer.

It was shown then that the transmissivity of the formation, on the points where it was known by pumping tests, was with a good approximation a linear function of this corrected transverse resistance. This permits to compute a transmissivity map from the transverse resistance, shown on Figure 7; although not identical, the main features of Figure 6 are reproduced. These different techniques may give clues to the interpretation of the structure of the basin, and check the result of one another, or draw the attention to particular areas where more geological data should be collected to confirm an interpretation.

CONCLUSIONS

To get a better evaluation of the structure and the properties of sediments, in a hydrologic context, it is clear that all available techniques should be used conjunctively. Four of them have been mentioned here: inverse problem from hydrologic data, geoelectricity, lithostratigraphic studies and detailed knowledge of the history of sedimentation, eventually through sedimentation models (see Jacod and Joathon, 1971; Joathon, 1971). They all give an element of interpretation of the geological context taken from different standpoints. Several others also may be considered such as other geophysical techniques, use of radioactive tracers, infrared airborne surveys, etc. However, geological control seems always to be necessary to check this type of results. They are nothing but links for the geological synthesis.

I should like to make a last comment concerning aquifer simulation, along the line initiated by Matheron (1965, 1969) and Kisiel (1971 and per. comm.). It deals with the precision of the previous evaluations. All the previously mentioned data are subject to errors, which makes them, in a stochastic context, random variables having expectation and variance. Variance of the initial conditions on one hand, of the controlled and uncontrolled input on the other hand, associated with variance of the parameters of the operators of the model (diffusivity equation), doubtlessly generates greater variance in the previsions of the model. These previsions can be physical (heads in the aquifer, for instance), or more generally economical (cost or benefits of the exploitation of the system) or even decisional (decisions to be made for the management of the system, including economical and sociological factors such as the recreational possibilities of a reservoir).

The aim should be to estimate the variance of these previsions, and to guide the data collection as well as the calibration procedure of the model towards minimization of this variance. This is far from being done.

If this seems a pessimistic point of view of the present state of the art, I shall give a last result. At the end of the work on the Aquitaine model, a well was drilled in the sands of the lower Eocene, where major modifications of transmissivities had to be made to calibrate the model. A transmissivity determination through a pumping test gave a result in excellent agreement with the modified values given to the model. Was this just good luck?

REFERENCES

Baetz, C., Gouvernet, C., Marsily, G. de, Potie, L., 1967, Comportement hydraulique d'un materiel alluvial heterogene. Vallee de la Mole (Var): Chronique d'Hydrogeologie, Edition BRGM, no. 12, p. 153-163.

BRGM, 1965, Carte piezometrique de la nappe captive des sables Eocenes en Gironde: BRGM, Ser. Geol. Reg. Aquitaine, chart.

Coats, K. H., Dempsey, J. R., and Henderson, J. H., 1970, A new technique for determining reservoir description from field performance data: Soc. Petroleum Eng. Jour., p. 66-74.

Duprat, A., Simler, L., Ungemach, P., 1969, Contribution de la prospection electrique a la recherche des caracteristiques hydrodynamiques d'un milieu aquifere: Terres & Eaux, v. 23, no. 62, p. 23-31.

Emsellem, Y., 1970, Construction de modeles mathematiques en hydrogeologie: Publ. Lab. d'Hydrogeologie Mathematique, Ecole des Mines, Fontainebleau, 129 p.

Emsellem, Y., Lerolle, Y., and Marsily, G. de, 1970, Identification automatique des parametres hydrauliques: Convegno Intern. sulle Acque Sotterranee, Palermo, 35 p.

Emsellem, Y., and Marsily, G. de, 1969, Restitution automatique des permeabilites d'une nappe. Le probleme inverse et la deconvolution: La Houille Blanche, no. 8, p. 861-868.

Emsellem, Y., Marsily, G. de, 1971, An automatic solution for the inverse problem: Water Res. Res., v. 7, no. 5, p. 1264-1283.

Emsellem, Y., Marsily, G. de, Poitrinal, D., Ratsimiebo, M., 1971, Deconvolution et identification automatique de parametres en hydrologie: IASH, Intern. Sym. on Math. Models in Hydrology, Warsaw, 32 p.

Freeze, R. A., 1971, Three dimensional, transient, saturated, unsaturated flow in a groundwater basin: Water Res. Res., v. 7, no. 2, p. 347-366.

Harbaugh, J. W., and Merriam, D. F., 1968, Computer applications in stratigraphic analysis: John Wiley & Sons, New York, 282 p.

Jacod, J., and Joathon, P., 1971, Use of random-genetic models in the study of sedimentary processes: Jour. Intern. Assoc. Math. Geology, v. 3, p. 265-280.

Jacquard, P., and Jain, C., 1965, Recherche sur l'interpretation des mesures de pression: Communication no. 19, Colloque de l'Association de Recherche sur les Techniques de Forage et de Production, Institut Francais du Petrole (Ref. IFP 11P12), 45 p.

Joathon, P., 1971, Representation d'une serie sedimentaire par un modele probabiliste. Application au processus d'ambarzoumian et etude de la structure de Chemery: These d'ingenieur-docteur, Nancy Univ. (no. CNRS AO 5805), 84 p.

Kisiel, C. C., 1971, General report on topic 7, objective functions and constraints in water resource systems: Session 15, Intern. Sym. on Math. Models in Hydrology, Intern. Assoc. Scientific Hydrology, Warsaw, 13 p.

Kleinecke, D., 1971, Use of linear programming for estimating geohydrologic parameters of groundwater basins: Water Res. Res., v. 7, no. 2, p. 367-376.

Marionnaud, J. M., 1967, Coupes lithostratigraphiques interpretatives dans le Tertiaire Nord Aquitain: BRGM, Ser. Geol. Reg. d'Aquitaine, 2 plates.

Marsily, G. de, 1972 , Chapitre "Calculs" de "Traite d'Informatique Geologique" A paraitre prochainement chez Masson, Paris, 40 p.

Matalas, N. C., 1969, System analysis in water resources investigations, in Computer applications in the earth sciences: Plenum Press, New York, p. 143-160.

Matheron, G., 1965, Les variables regionalisees et leur estimation: Masson et Cie, Paris, 305 p.

Matheron, G., 1967, Elements pour une theorie des milieux poreux: Masson et Cie, Paris, 166 p.

Matheron, G., 1969, Le krigeage universel: Cahiers du Centre de Morphologie Mathematique (Fasc. 1), Ecole des Mines, Fontainebleau, 83 p.

Matheron, G., and Huijbrechts, C., 1970, Universal kriging; an optimal method for estimating and contouring in trend surface analysis: Intern. Sym. Tech. for Decision Making in the Min. Ind., CIM, Montreal, 15 p.

Nelson, W., 1970, A computer system for the analysis of flow and water quality in large heterogeneous groundwater basins: Convegno Intern. sulle Acque Sotteranee, Palermo, 9 p.

Prickett, T. A., and Lunnqvist, C. G., 1968, Comparison between analog and digital simulation for aquifer evaluation, in The use of analog and digital computer in hydrology: Tucson, Publ. IASH, Tucson, no. 81, p. 625-634.

Remson, I., and others, 1971, Numerical methods in subsurface hydrology, with an introduction to the finite elements method: John Wiley & Sons, New York, 330 p.

Schoeller, H., and others, 1969, Synthese des etudes sur le systeme multicouche des nappes tertiaires du nord de l'Aquitaine: La Houille Blanche, no. 8, p. 907-918.

Schoeller, H., and others, 1971, Etude par modele mathematique des risques de pollution par les eaux salees des aquiferes souterrains de Gironde: Intern. Union of Geodesy & Geophysics, 15th Gen. Assembly (and IASH), Moscow, 25 p.

Ungemach, P., Mostaghimi, F., and Duprat, A., 1969, Essais de determination du coefficient d'emmagasinement en nappe libre. Application a la nappe alluviale du Rhin: Intern. Assoc. of Scientific Hydrology, 14th annee, no. 2, p. 169-190.

Weber, E. M., Peters, H. J., and Frankel, M. L., 1968, California's digital computer approach to ground water basin management studies, *in* The use of analog and digital computers in hydrology: Publ. IASH, Tucson, no. 80, p. 215-223.

Witherspoon, P. A., Javandel, I., and Newman, S. P., 1968, Use of finite elements method in solving transient flow problems in aquifer systems - *in* The use of analog and digital computer in hydrology: Publ. IASH, Tucson, no. 81, p. 687-698.

Chateau
Leoville Poyferre
1924

APPENDIX

LES CRUS CLASSES DU MEDOC

(classified vineyards of Medoc)

- Classification of 1855 -

CHATEAUX	LOCALITY	PRODUCTION (Tonneaux)
	Premiers crus	
Lafite	Pauillac	180
Latour	Pauillac	100
Margaux	Margaux	150
Haut-Brion Graves	Pessac	100
	Deuxiemes crus	
Mouton-Rotschild	Pauillac	95
Rausan-Segla	Margaux	60
Rauzan-Gassies	Margaux	50
Leoville-Las-Cases	Saint-Julien	150
Leoville-Poyferre	Saint-Julien	120
Leoville-Barton	Saint-Julien	100
Durfort-Vivens	Margaux	30
Lascombes	Margaux	35
Gruaud-Larose	Saint-Julien	185
Brane-Cantenac	Cantenac	100
Pichon-Longueville	Pauillac	78
Pichon-Longueville-Lalande	Pauillac	100
Ducru-Beaucaillou	Saint-Julien	130
Cos-d'Estournel	Saint-Estephe	50
Montrose	Saint-Estephe	100
	Troisiemes crus	
Kirwan	Cantenac	100
Issan	Cantenac	30
Lagrange	Saint-Julien	100
Langoa	Saint-Julien	75
Giscours	Labarde	20
Malescot-St Exupery	Margaux	50
Cantenac-Brown	Cantenac	90

CHATEAUX	LOCALITY	PRODUCTION (Tonneaux)
Palmer	Cantenac	100
La Lagune	Ludon	80
Desmirail	Margaux	30
Calon-Segur	Saint-Estephe	150
Ferriere	Margaux	20
d'Alesme-Becker	Margaux	20
Boyd-Cantenac	Cantenac	30
	Quatriemes crus	
Saint-Pierre-Sevaistre	Saint-Julien	40
Saint-Pierre-Bontemps	Saint-Julien	60
Branaire-Ducru	Saint-Julien	100
Talbot	Saint-Julien	140
Duhart-Milon	Pauillac	140
Pouget	Cantenac	30
La Tour-Carnet	Saint-Laurent	70
Rochet	Saint-Estephe	60
Beychevelle	Saint-Julien	100
Le Prieure-Lichine	Cantenac	30
Marquis-de-Terme	Margaux	75
	Cinquiemes crus	
Pontet-Canet	Pauillac	200
Batailley	Pauillac	80
Hart-Batailley	Pauillac	40
Grand-Puy-Lacoste	Pauillac	70
Grand-Puy-Ducasse	Pauillac	35
Lynch-Bages	Pauillac	100
Dauzac	Labarde	60
Mouton-d'Armailhacq	Pauillac	100
Le Tertre	Arsac	100
Pedesclaux	Pauillac	30
Belgrave	Saint-Laurent	150
Camensac	Saint-Laurent	70
Cos-Labory	Saint-Estephe	45
Clerc-Milon-Mondon	Pauillac	35
Croizet-Bages	Pauillac	50
Cantemerle	Macau	100

MATHEMATICAL SEARCH PROCEDURES IN FACIES MODELING IN SEDIMENTARY ROCKS

F. Demirmen

N. V. Turkse Shell

ABSTRACT

Lack of a standard or generally accepted method of mathematically defining facies from a set of attributes in sedimentary rocks makes a priori selection of a particular method difficult. It is proposed that different combinations of alternative classificatory techniques, leading to competitive classifications, be used to arrive at the first approximations of the facies model. An assessment of these models then can be made, and the "best" one selected further improved on the basis of intuitively appealing objective criteria. These search procedures relieve the investigator of the rigidity of a particular method and also place his empirical facies model on a more rational basis in retrospect, allowing more meaningful interpretations. These principles are illustrated in a facies study of Pennsylvanian carbonate rocks (from southeastern Utah) in which principal component analysis, Euclidean distance function, standardization, and weighted and unweighted pair-group methods, are among the classificatory schemes employed in facies construction.

INTRODUCTION

In recent years geologists have been increasingly interested in classifying sedimentary rocks on the basis of a set of measurements or observation values in an attempt to recognize facies in these rocks. The resulting classifications, although they lack the usual attributes of deterministic, stochastic, or other formal statistical models, can nonetheless be regarded as empirical facies models in that they purport to depict natural groupings in rocks and provide a framework for interpreting the depositional environment. With the insight they provide into the group structure in the sample space and into depositional phenomena, these empirical models may lead, at a more advanced stage of analysis, to the construction of more formal models, although this need not be so. Empirical facies models usually have enough predictive and interpretive value that they may be an end in themselves.

When we attempt to construct facies models from a given set of attributes, we immediately run into the problem of choosing among the techniques available. A whole gamut of mathematical techniques, ranging from the transformation of the original data matrix to final clustering, has been designed for defining facies or, more generally, for classifying items (e.g. Sokal and Sneath, 1963; Feldhausen, 1970; Merriam, 1970; Switzer, 1970; Anderson, 1971). Yet, we lack objective criteria to decide, a priori, which of these techniques is better, although certain techniques have been favored in empirical facies modeling, more by convention than on rational basis. The diversity of techniques usually leads to a diversity of end-products. This implies that, starting from the same data matrix, but using different data manipulations, it is possible to arrive at different facies models.

Another and allied problem suggests itself at this point. If we have different facies models from the same basic data, how do we reconcile them or make a selection among them? The ultimate test of the relative validities of empirical facies models must be made on substantive grounds; of two models, the one that makes more sense geologically or the one that leads to a more meaningful environmental interpretation can be considered more valid. Such an assessment, however, would be highly subjective, and it seems best to use this procedure of testing as the last resort. Instead, the use of objective criteria to test the relative merits of our facies models seems to have great appeal. Furthermore, it would be advantageous if a given facies model were further improved.

To alleviate the difficulties raised above, it is proposed that search procedures be used to sort out the most suitable empirical facies model. The philosophy is followed that, instead of adhering to a particular method, the investigator should allow himself a latitude in the choice of methods, make use of competitive techniques, and assess the results in retrospect to arrive at a meaningful facies model. Subjective judgment should be kept to a minimum and used as the last resort. The basic steps involved are as follows.

(1) Use different combinations of alternative classificatory schemes to arrive at classifications which will, in general, be different. These classifications represent first approximations

to our facies model. This approach is analogous to the shotgun method advocated by Miller and Kahn (1962, p. 315) for analyzing geologic data in the absence of a priori hypotheses. Because it is impractical to use all the existing classificatory schemes, the investigator would be justified to employ the combination of techniques which he prefers.

(2) Assess, a posteriori, the relative qualities of the preliminary facies models obtained above, and select the "best". The criteria of Demirmen (1969) for evaluating classifications can be used for the assessment here, although other methods or criteria of assessment may also be appropriate. The approach here is analogous to testing of the statistical hypotheses, but without the rigor and probabilistic connotations of that method.

(3) Improve further, if possible, the best model obtained, by a suitable criterion such as the one given by Demirmen (1969).

These search procedures were used in a facies study of Pennsylvanian Honaker Trail Formation, exposed and sampled along the San Juan River in southeastern Utah (USA). The application of the search procedures to these rocks is described here.

RAW DATA

Our raw data consist of estimates of percentage and indices of coarseness obtained over a suite of 200 thin sections representing the Honaker Trail limestones. Altogether 14 variables are used, 12 of them being percentages and 2 of them coarseness indices. The 12 constituents of which percentages were estimated are, in abbreviated and coded form: FORS (foraminifers + osagia), BRYO (bryozoans), BRCH (brachiopods), MOLL (mollusks), ECHI (echinoderms), TRIL (trilobites), OSTR (ostracodes), PELI (pellets + intraclasts), OOID (ooids), MUDN (matrix microcalcite + silt-sized or finer calcareous debris), ORTH (matrix orthospar, i.e., primary crystalline spar occurring in matrix), and TERR (terrigenous material) (see also Demirmen, 1971).

The category of undifferentiated and unknown constituents was not included among the percentage variables because this category has a dubious interpretive value. Elimination of this category also tends to open the data and reduce the distortion of correlations between percentage variables in a closed data set (Herdan, 1953, p. 296-304; Chayes, 1962) in which the percentages add to 100.

Coarseness index (CI) is a measure of particle coarseness in a thin section (Demirmen, 1971). This index was measured for allochem particles (CI-A) and terrigenous particles (CI-T) separately, yielding two additional variables.

The 14 variables used for facies construction can be regarded as containing the maximum amount of quantifiable meaningful petrographic information that characterizes each thin section. Inasmuch as the ultimate aim of facies construction is to interpret the environment of deposition, the attributes that pertain to post-depositional alterations are not included among the variables, although such information would be useful and was used in the final environmental analysis.

The summary of our raw data, listing the means, standard deviations, and coefficients of variation, is given in Table 1.

Table 1. - Summary of raw data.

	FORS	BRYO	BRCH	MOLL	ECHI	TRIL	OSTR	PELI	OOID	MUDN	ORTH	TERR	CI-A	CI-T
Means	12.0	5.4	3.2	8.1	12.9	0.4	0.2	1.1	3.6	39.0	3.8	9.0	1.6	0.2
Standard deviations	11.0	6.3	3.7	8.4	14.3	0.5	0.5	4.8	11.8	23.0	8.0	12.1	0.9	0.1
Coefficients of variation	0.92	1.16	1.14	1.04	1.11	1.20	2.20	4.34	3.27	0.59	2.12	1.35	0.56	0.41

It is seen that MUDN, with a mean of about 39 percent, is the most abundant constituent in our raw data, followed next by ECHI (13 percent) and FORS (12 percent). TRIL and OSTR are the least abundant constituents, each with a mean of less than 1 percent. ORTH is less than 4 percent, indicating the general scarcity of sparites in the samples. The allochems generally are coarser than the terrigenous particles. The greatest absolute variation (standard deviation) is found in MUDN, which is not surprising because this is the most abundant constituent. Note, however, that the greatest variation relative to the mean (coefficient of variation) occurs with PELI and OOID, with OSTR and ORTH trailing behind.

We now designate our raw data by the matrix **F** (nxp) whose element f_{ki} represents the measurement of the k-th item (thin section) on the i-th variable, i.e., $\mathbf{F} = (f_{ki})$, $k = 1,\ldots, n$; $i = 1,\ldots, p$. Thus, k-th row vector of $\mathbf{F}$ represents the k-th item in a p-dimensional Euclidean space identified by the f-variables. For our data, $n = 200$ and $p = 14$.

STANDARDIZATION OF RAW DATA

Our raw data include two different types of variables: percentages and coarseness indices. These are measures of different attributes and therefore not comparable with each other. Principal components have little meaning if the original variables from which they are generated measure different attributes. Furthermore, the Euclidean distances, which we use as a measure of similarity between our thin sections, would be greatly influenced by those variables having a large variance and only slightly influenced by variables having a small variance if this function is computed directly from our raw scores. Biasing the similarities according to variance in this manner may be objectionable because variables with small variance may contain valuable information for classification and there is no a priori reason for underrating their importance. For these reasons, we standardize our original f-variables to a new set of z-variables and obtain a new score matrix $\mathbf{Z}$ $(n\mathrm{x}p)$:

$$Z = (z_{ki}),\ z_{ki} = \frac{f_{ki} - f_{.i}}{s_i},\ k = 1,\ldots, n;\ i = 1,\ldots, p, \qquad (1)$$

where $f_{.i}$ and s_i are the mean and standard deviation, respectively, of i-th f-variable computed over $n = 200$ items. Clearly, each z-variable has mean 0 and variance 1, "measured" in terms of the standard deviation of the corresponding original variable. Thus the z-variables are comparable with each other in mean, variance, and "units of measurement". If we define a matrix $\bar{\mathbf{F}}$ $(n\mathrm{x}p)$such that the i-th column of $\bar{\mathbf{F}}$ contains uniformly the mean $f_{.i}$, an alternative expression of (1) is:

$$Z = (F - \bar{F})\ D^{-1}(s_i). \qquad (2)$$

Here $D^{-1}(s_i)$ is the inverse of the diagonal matrix whose principal diagonal contains the standard deviations s_i arranged in descending order according to the subscript $i = 1,\ldots, p$. Note that the $\mathbf{Z}$ matrix, like the $\bar{\mathbf{F}}$ matrix, is of order (200 x 14).

PRINCIPAL COMPONENTS

Although our standardized z-variables have the same mean and variance, they are nevertheless correlated. Their correlation is a carryover from the original f-variables. In fact, it is easy to see that, if the correlation between any two of the original variables f_i and f_j is r_{ij}, then the correlation between the standardized variables z_i and z_j derived from f_i and f_j is also r_{ij}. Correlated variables can be regarded as containing redundant information, and their use in the computation of the similarity measures is best avoided. To accomplish this, we transform the z-variables into a set of principal components, which we designate as the g-variables:

$$G = Z\,A. \tag{3}$$

Here $A(p\text{x}p)$ is an orthogonal matrix whose columns represent the normalized eigenvectors of the correlation matrix $R = 1/(n-1)Z'Z$ computed in the z-space (also in the f-space). Accordingly, $G(n\text{x}p)$ is our new score matrix whose i-th column contains the scores of n items with respect to the i-th principal component g_i. Principal components obtained in this manner are mutually uncorrelated, with the i-th principal component having a variance equal to the i-th largest eigenvalue of R, which we designate as λ_i. Furthermore, the total variance in the z-space is identical to the total variance in the g-space, i.e. $p = \Sigma\lambda_i$. The proportion of total variance attributable to g_i is λ_i/p, and the proportion of cumulative variance associated with g_i is $\Sigma_{j=1}^{i}\lambda_j/p\,(j\leq i)$. Thus the principal components, whereas preserving the total variance in the z-space, account for successively smaller portions of the total variance as their index (subscript) i approaches p. The first principal component, for example, has the largest variance, the second principal component the second largest variance, etc. This property of the principal components allows us to reduce the dimensionality of the data without sacrificing much variance or information. We can eliminate the last $t(<p)$ principal components and retain a large portion of the total variance. It is possible, of course, that some of the principal components have zero variance, in which situation they can be readily ignored. This would happen if R is of rank less than p.

A measure that relates the principal components to the original f-variables is their product-moment correlations. These correlations have a simple relation to the eigenvalues and eigenvectors associated with the principal components. If we let $Q(p\text{x}p)$ be a square matrix whose q_{ij} is the correlation between f_i and g_j, then it can be readily shown that

$$Q = A\,D\,(\sqrt{\lambda_i}), \tag{4}$$

where $D(\sqrt{\lambda_i})$ is a diagonal matrix containing the elements $\sqrt{\lambda_i}$ $(i = 1,\ldots, p)$ in its principal diagonal. Clearly, equation (4) implies that

$$q_{ij} = a_{ij} \sqrt{\lambda_j} \quad \text{for all } i, j = 1,\ldots, p. \tag{5}$$

From the properties of the product-moment correlation, q_{ij} also represents the correlation between z_i and g_j. Note that, unlike R, Q is in general nonsymmetric.

Two additional aspects are noteworthy here. First, equation (3) implies that

$$g_i = \sum_{j=1}^{p} z_j a_{ji} \quad \text{for all } i = 1,\ldots, p, \tag{6}$$

i.e., each principal component g_i is a linear combination of the z-variables, with the elements of the i-th column of A serving as the coefficients of linearity. If a principal component is to measure anything meaningful, the variables from which it is derived must be comparable. This explains why we used the z-variables rather than the original f-variables to obtain our principal components. Second, the transformation from the z-space to the g-space (eq. 3) is an orthogonal transformation, so that the ordinary Euclidean distance computed between any pair of items in the g-space is identical to the corresponding Euclidean distance in the z-space. If the variances of the principal components are equalized, however, this relationship does not hold.

The principal component results of our data are summarized in Tables 2 and 3. Table 2 gives the variance information about the principal components.

Table 2. - Eigenvalues associated with principal components, and percentage and cumulative percentage of total variance attributed to each principal component.

Eigenvalues													
1	2	3	4	5	6	7	8	9	10	11	12	13	14
3.525	3.001	1.482	1.171	0.950	0.803	0.765	0.593	0.487	0.386	0.357	0.268	0.210	0.001
Percentage of total variance													
1	2	3	4	5	6	7	8	9	10	11	12	13	14
25.2	21.4	10.6	8.4	6.8	5.7	5.5	4.2	3.5	2.8	2.6	1.9	1.5	0.0
Cumulative percentage of total variance													
1	2	3	4	5	6	7	8	9	10	11	12	13	14
25.2	46.6	57.2	65.6	72.4	78.1	83.6	87.8	91.3	94.0	96.6	98.5	100.0	100.0

Table 3. - Product-moment correlations between original variables and first three principal components.

	1	2	3
FORS	-0.186	-0.764	-0.323
BRYO	-0.602	0.259	-0.076
BRCH	-0.728	0.309	0.035
MOLL	-0.180	-0.756	-0.277
ECHI	-0.783	0.236	0.071
TRIL	-0.734	0.069	-0.101
OSTR	0.130	0.113	-0.138
PELI	0.133	-0.245	0.605
OOID	0.271	-0.153	0.814
MUDN	0.416	0.753	-0.150
ORTH	0.070	-0.812	0.131
TERR	0.568	0.045	-0.350
CI-A	-0.804	0.122	0.257
CI-T	-0.399	-0.534	-0.065

Table 3 lists the correlations between all the original variables and the first three principal components. We note from these tables that the first three principal components account for 57 percent of the total variance, whereas the first 10 components account for 94 percent. The first component, containing about 25 percent of the total variance, identifies a dimension where the samples with high values of CI-A, ECHI, TRIL, BRCH, and BRYO tend to be separated from those with low values in these variables. High positive scores on the first component denote relatively low z-values in one or more of these variables, and vice versa. The second component, with 21 percent of the total variance, tends to differentiate the samples on the basis of their ORTH, FORS, MOLL, and CI-T values, with high scores on this component denoting relatively low z-values in one or more of these variables, and vice versa. This component has a relatively high correlation with MUDN. The third component, with only 11 percent of the total variance, ranks the samples according to their OOID and PELI contents, with high positive scores on this component corresponding to high z-values in one or both of these variables. Similar arguments can be given for the remaining, less important principal components.

SIMILARITY MEASURES

The next step in our facies construction is the computation of similarity measures which serve as the basis of grouping of our samples. The similarity measures are computed between each pair of samples. In general, for n samples, $\binom{n}{2} = n(n-1)/2$ similarity measures are required.

Basic Concept

The measure of similarity we use is based on the concept of ordinary Euclidean distance. This means that we visualize the n items as a swarm of n points in a Euclidean space identified by p Cartesian coordinate axes. Each item in this space is represented by a point whose coordinates are the measurements of this item with respect to the p variables. We use the distance between two items as a measure of their resemblance, with small distances corresponding to high similarities, and vice versa. In general terms, if our measurements are defined by x-variables, and if we let x_{hi} be the measurement of the h-th item on the i-th variable $(i=1,\ldots, p)$, and similarly for x_{ki}, the measure of similarity, designated as $\delta(h,k)_{(x)}$, between the h-th item and the k-th item is given by

$$\delta(h,k)_{(x)} = c - \sqrt{\frac{1}{p} \sum_{i=1}^{p} (x_{hi} - x_{ki})^2} , \tag{7}$$

where c is some arbitrary positive constant larger than the largest Euclidean distance. To avoid too many significant figures after the decimal, c should be chosen to be a round number not too different from the largest Euclidean distance. Hereafter, we shall call $\delta(h,k)_{(x)}$ the distance coefficient between the h-th and k-th items in the x-space.

Options

Having decided on the similarity function, our next concern is to define the parameters in (7). The selection of c of course is no problem, and we arbitrarily set $c = 10$. The designation of the type and number of the x-variables, however, is not so straightforward, and a number of alternatives come to mind. One obvious possibility is to let the x-variables be the principal components, i.e. set $x_i = g_i$ for all $i = 1,\ldots, p$. But do we need to use all the principal components, i.e., set $p = 14$? As stated earlier, the ordinary Euclidean distances in the principal component space, if all the principal components are retained, are identical to the corresponding Euclidean distances in the standardized z-space, so that setting $x_i = g_i$ and $p = 14$ is tantamount to computing the distance coefficients in the z-space. On the other hand, because the last principal components contribute little to the total variability, we may wish to eliminate these components from the computation, i.e., set p to some number less than 14. The distance coefficients computed in this manner would not be too different from their counterparts corresponding to the situation $p = 14$; we would, however, achieve parsimony in the dimensionality of our space.

Two other alternatives that come to mind concern standardization. We can think of component-wise standardization, whereby the variances of the principal components, as measured over n items, are equalized, e.g., set to 1 (the means computed over n items are already equal, i.e., 0). In this manner, we would not weight the principal components according to their variances in the computation of the distance coefficients, a practice which may be objectionable because it would automatically and a priori underrate the information value of those principal components having small variance. Component-wise standardization would not affect the orthogonality of the principal components. Alternatively, we can think of item-wise standardization, whereby the means and variances of the component scores, as measured over p principal components, are equalized, say set to 0 and 1, respectively. These means and variances then become comparable from item to item. It can be shown that distance coefficients computed after item-wise standardization have a simple relation to item-wise correlation coefficients, another similarity measure used in some problems of classification.

It is clear from the foregoing that there are a number of alternatives for defining the variables on which to base our distance function. With each alternative the distance coefficients, and hence the resulting classifications, will in general be different. At present, we have no a priori rational basis for choosing among these alternatives. We propose to approach this problem empirically: We try a number of options and then assess the results in retrospect.

Option 1. - Standardize the principal components component-wise and retain all 14 of them. In this situation we define a new score matrix $G^{(c)}$ (200 x 14),

$$G^{(c)} = G\, D^{-1}\ (\sqrt{\lambda_i}),\ i = 1, \ldots, 14, \tag{8}$$

and compute the distance coefficients from these new scores. Thus, returning to (7), we set $c = 10$, $p = 14$, and $x_{ki} = g_{ki}^{(c)}$, where

$$g_{ki}^{(c)} = \frac{g_{ki}}{\sqrt{\lambda_i}} \tag{9}$$

for all $k = 1, \ldots, 200$; $i = 1, \ldots, 14$. Clearly, the transformation (8) equalizes the variances of the principal components and sets them to 1. The new $g^{(c)}$ variables are uncorrelated, each with zero mean and unit variance.

Option 2. - Use the principal components, as they are, but retain only the first ten. This option is equivalent to setting $c = 10$, $p = 10$, and $x_{ki} = g_{ki}$ for all $k = 1, \ldots, 200$, and $i = 1, \ldots, 10$ in (7). We recall that the first ten principal components account for 94 percent of the total variance.

Option 3. - Take the first 10 principal components and standardize them sample-wise. In this situation we define a new score matrix $G^{(s)}$ (200 x 10) such that

$$G^{(s)} = \left(g_{ki}^{(s)}\right) = \left(\frac{g_{ki} - g_{k.}}{s_{(k)}}\right), \quad \begin{array}{l} k = 1, \ldots, 200; \\ i = 1, \ldots, 10, \end{array}$$

where (10)

$$g_{k.} = \frac{1}{10} \sum_{i=1}^{10} g_{ki}, \text{ and } s_{(k)} = \sqrt{\frac{1}{3} \sum_{i=1}^{10} (g_{ki} - g_{k.})^2} .$$

We then compute the distance coefficients in the $g^{(s)}$ space, setting $c = 10$, $p = 10$, and $x_{ki} = g_{ki}^{(s)}$ for all $k = 1, \ldots, 200$, and $i = 1, \ldots, 10$ in (7). The transformation (10) sets the means and variances of all items to 0 and 1, respectively.

GROUPING METHODS

After the computation of the similarity measures, we proceed to group the $n = 200$ items or samples on the basis of these similarities. A number of techniques have been designed for the purpose, but the two schemes we use are the so-called weighted and unweighted pair-group methods. First developed and advocated by biologists, these two methods have found extensive use in geology. A good description of these methods is given in Sokal and Sneath (1963). The rationale behind using both of these methods rather than just one is that we do not know a priori which of these methods is better. A few attempts on the part of biologists to assess these two methods have given inconclusive results, although the unweighted scheme seems to better condense the initial similarity matrix (Sokal and Sneath, 1963, p. 191). As in the situation of the similarity measure, our approach to the problem of selecting between the two grouping methods is empirical. We try both methods and evaluate the results in retrospect.

COMBINATION TRIAL PROCEDURES

Because we have three options on the distance coefficients and two options on the grouping methods, we have at our disposal six alternative schemes of classifying our data. These schemes represent trial or search methods devised to arrive at an empirical facies model from our data. For convenience, we call each combination of a distance coefficient and a grouping method a Procedure. We designate these combinations as "1-A", "1-B", "2-A", "2-B", "3-A", and "3-B", with the numeral referring to the option of distance coefficient and the letters A and B referring to the unweighted and weighted pair-group methods, respectively. More specifically, we have:

Procedure 1-A: Combination of Option 1 distance coefficient and the unweighted method; i.e., we standardize the principal components component-wise, retain all 14 of them, and apply the unweighted method.

Procedure 1-B: Combination of Option 1 distance coefficient and the weighted method; i.e., the same as above, but we use the weighted method.

Procedure 2-A: Combination of Option 2 distance coefficient and the unweighted method; i.e., we use the first 10 principal components without standardization, and apply the unweighted method.

Procedure 2-B: Combination of Option 2 distance coefficient and the weighted method; i.e., the same as 2-A, but we use the weighted method.

Procedure 3-A: Combination of Option 3 distance coefficient and the unweighted method; i.e., we take the first 10 principal components, standardize them item-wise, and apply the unweighted method.

Procedure 3-B: Combination of Option 3 distance coefficient and the weighted method; i.e., the same as 3-A, but we use the weighted method.

DENDROGRAMS

The results obtained from the application of the six combination Procedures are portrayed in dendrogram form in Figures 1, 2 and 3 (shadings to be ignored for the moment). Each dendrogram in effect depicts the manner the individual items (samples) and the larger clusters generated from them were grouped during the clus-

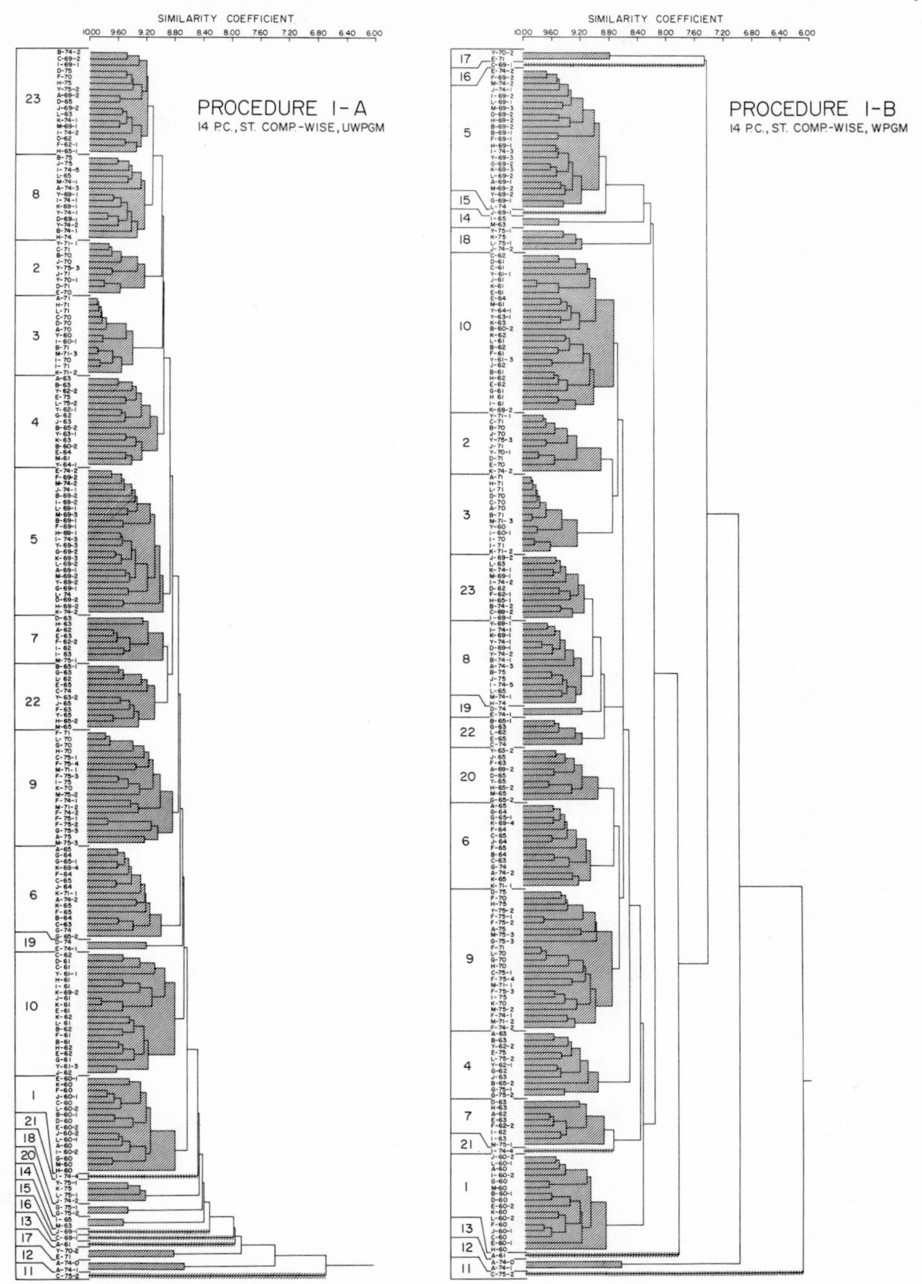

Figure 1. - Dendrograms and preliminary classifications obtained by combination Procedures 1-A and 1-B. Numbers (1 through 23) on left margin are class indices.

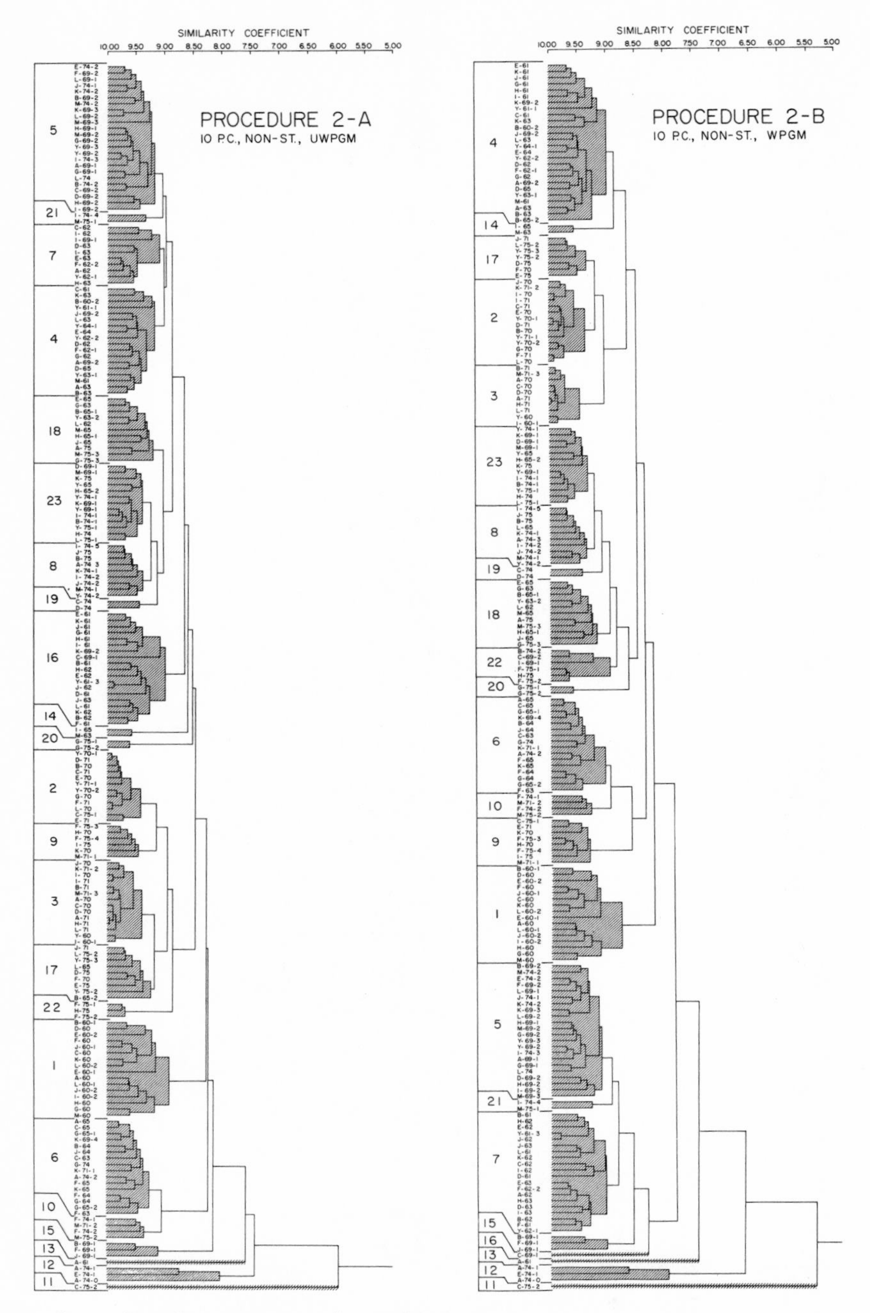

Figure 2. - Dendrograms and preliminary classifications obtained by combination Procedures 2-A and 2-B. Numbers (1 through 23) on left margin are class indices.

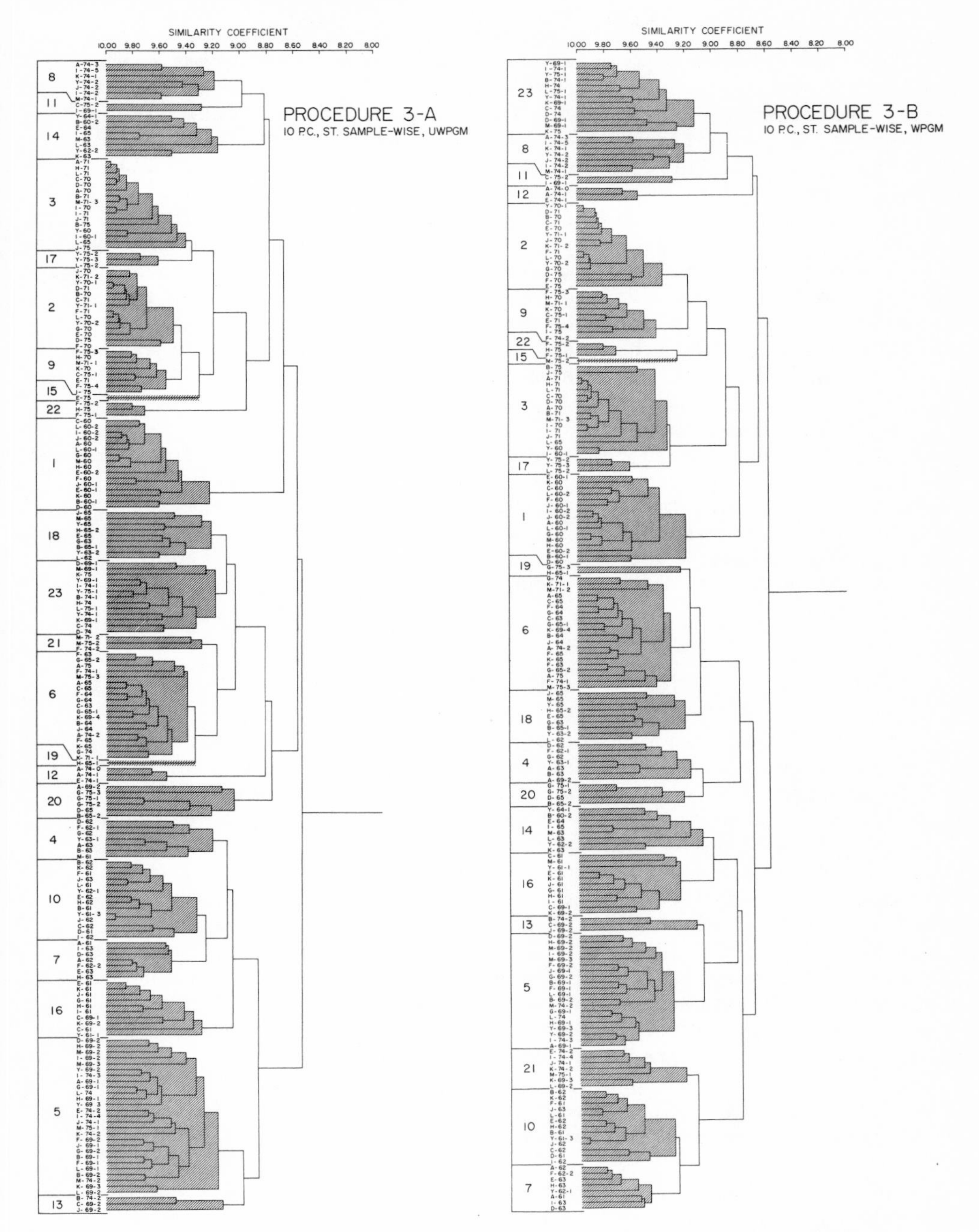

Figure 3. - Dendrograms and preliminary classifications obtained by combination Procedures 3-A and 3-B. Numbers (1 through 23) on left margin are class indices.

tering cycles in the weighted and unweighted pair-group methods. The items are listed on the left margins of the dendrograms. The similarity value at which any two items or clusters were joined during the clustering process is indicated by the vertical line that connects the horizontal stems representing these items or clusters. The similarity value corresponding to a vertical line is read from the horizontal axis on which the similarities are scaled. Because the values of the similarity measures differ according to the distance coefficient option used, the similarity values are not comparable from one figure to another, although for the two dendrograms in a given figure, the similarities are comparable. Note, however, that in each dendrogram the highest possible value is 10. This is due to the fact that we have used c = 10 in all three of our distance coefficient options.

A hierarchical grouping, with the clusters becoming progressively larger toward the right, is evident in the dendrograms. The last horizontal stem in a dendrogram represents all the items in that dendrogram. This hierarchic structure is characteristic of the pair-group methods in general. A comparison of the dendrograms reveals significant differences between them. In general, the discrepancies are more pronounced at the lower levels of similarity.

DEFINITION OF PRELIMINARY MODELS

The principal use of dendrograms is that they allow us to classify our data. When we attempt to do this, however, there arises the problem of recognizing or picking the classes. An inspection (without regard to shadings) of Figures 1, 2, and 3 reveals that we have a wide spectrum of choice for picking classes from a dendrogram, depending on what we mean by a class and how many classes we wish to select. An argument that the classes defined should be "natural" is of little help because a natural class is difficult to define. The common practice among biologists and geologists has been to draw a vertical line, called the phenon line, at an arbitrary level of similarity measure and pick the classes as those clusters whose stems are intersected by this line. We deviate from this practice for the following reasons:

(1) No convincing argument has been given that this method of picking classes is necessarily a good one.

(2) Delineation of classes by a phenon line would in some instances yield classes that are too large (containing a large proportion of the items). There would be a need to further partition these large classes.

(3) Moving the phenon line in one or the opposite direction would in some instances result in a classification that is highly different from the existing one. A classification should not be drastically influenced by such slight translations of a line which is of questionable value in the first place.

The approach we take to the problem of picking classes in a dendrogram is an intuitive one. We adopt the following principles:

(1) A class is that item or collection of items that, upon visual inspection of the dendrogram, tends to stand out from the neighboring items or clusters.

(2) A class should not contain a large proportion of the items. Large classes tend to include items of diverse character and defeat the purpose of classification. Note, however, that classes with only one or two items are permissible. This allowance is made in order to effectively separate those items with distinct properties, such as the sample C-75-2 in Figures 1, 2, and 3.

(3) The classes should be chosen in a manner that the classifications obtained from the six dendrograms are not too different from each other. The rationale behind this principle is as follows: No matter how we pick the classes, the classifications obtained from the six dendrograms are different. As will be seen in the sequel, we shall attempt to improve upon the classifications obtained from these dendrograms. It is reasonable to expect that the improved classifications will converge if the initial classifications on which they are based are not too different. Furthermore, our confidence in the initial classifications would be enhanced if these classifications did not differ from each other too much.

(4) The number of classes should be small enough to be of practical use. This principle is not alwasy easy to reconcile with principle (3). As noted before, the six dendrograms in general become more dissimilar as we proceed from high similarities to low similarities. Hence principle (3) in general necessitates the selection of classes at high levels of similarity, which means a relatively large number of classes. If principles (3) and (4) cannot be reconciled, we try to reach a compromise.

In addition to the guidelines noted above, we make the additional stipulation that the number of classes defined in all the six dendrograms will be the same. This constraint is necessary for the assessment step described next.

The classes defined under these considerations are indicated by shaded patterns in Figures 1, 2, and 3. In each dendrogram 23 classes, indexed 1 through 23, are delineated. The indices are assigned in such a manner that the classes with identical index numbers are, with some exceptions, comparable from one dendrogram to another. It will be noted that the resultant classifications, while different in detail, are by and large in agreement from one dendrogram to another. Some classifications are more similar than others. In particular, the classifications obtained using the same similarity option but different grouping methods (weighted vs. unweighted method) seem to be more similar to each other than the classifications obtained using the same grouping method but different similarity options.

The alternative classifications shown in Figures 1, 2, and 3 represent preliminary or trial facies models for our data. They are first approximations to our empirical facies model, and are the subject of further analysis. For convenience, we designate these preliminary models as 1-A, 1-B, etc., according to the Procedures by which they were obtained.

ASSESSMENT OF PRELIMINARY MODELS

The next step is the assessment of the preliminary classifications or facies models. To achieve this, we use the tr $\mathbf{W}$, Λ, and tr $\mathbf{W}^{-1}\mathbf{B}$ criteria of Demirmen (1969), which compares and evaluates classifications on the basis of their compactness. Of any two classifications, the one that is more compact by a given criterion can be regarded as better by that criterion. Small values of tr $\mathbf{W}$ and Λ, and large values of tr $\mathbf{W}^{-1}\mathbf{B}$ imply relatively good classifications. The use of these criteria requires that the classifications compared contain the same number of classes, hence the constraint noted.

The score matrix we use to compute the three criteria is some matrix Y (200 x 14), identical to the $G^{(c)}$ (200 x 14) matrix except for a constant factor of $1/\sqrt{n-1}$:

$$Y = \frac{1}{\sqrt{n-1}} G^{(c)}. \qquad (11)$$

For a particularly preliminary classification, the $\mathbf{Y}$ matrix then is partitioned (i.e., the rows are arranged into classes) so as to represent that classification. We recall that the $g^{(c)}$ variables are uncorrelated each with mean (grand mean) 0 and variance (grand variance) 1. The y-variables also are uncorrelated, each with mean 0, but with variance $1/(n-1)$. We thus utilize equal-mean, equal-variance uncorrelated variables to compare our preliminary models. The partitioning of the **Y** matrix in any way does not affect the means (grand means), the variances (grand variances), and the uncorrelatedness (as measured over all items). The division by $\sqrt{n-1}$ above is intended to merely reduce the scales of the $g^{(c)}$ variables to meet the output requirements of the computer program used for this purpose. The comparison of classifications by our criteria leads to identical conclusions whether we use the $g^{(c)}$ variables or the y-variables.

The computed values of our criteria for the six preliminary models and their relative ratings (rankings) are shown in Table 4.

Table 4. - Assessment of preliminary models obtained by six combination Procedures. For given criterion, rating 1 denotes best, rating 2 denotes second best, etc.

	PRELIMINARY					
	1-A	1-B	2-A	2-B	3-A	3-B
CRITERION						
tr $\underline{W}$	4.80	4.92	5.58	5.84	6.04	5.96
$\Lambda \times 10^9$	15.23	25.23	13.82	28.43	123.6	92.98
tr $\underline{W}^{-1}\underline{B}$	50.15	47.03	63.45	59.78	49.44	49.53
RATING						
tr $\underline{W}$	1	2	3	4	6	5
Λ	2	3	1	4	6	5
tr $\underline{W}^{-1}\underline{B}$	3	6	1	2	5	4

We note that the ratings by the tr **W** and Λ criteria are by and large in agreement whereas, relative to these criteria, the ratings by the tr $\mathbf{W}^{-1}\mathbf{B}$ criterion are somewhat erratic. Considering the three criteria simultaneously, we observe that the poorest ratings correspond to the preliminary models 3-A and 3-B. The preliminary models 1-B and 2-B have intermediate ratings, with 1-B rating slightly better. The best ratings correspond to the preliminary models 1-A and 2-A. The ratings of these models relative to each other, however, are not unequivocal. The preliminary model 1-A rates first (best) by the tr **W** criterion, second by the Λ criterion, and third by the tr $\mathbf{W}^{-1}\mathbf{B}$ criterion. By contrast, the preliminary model 2-A ranks first by the Λ and tr $\mathbf{W}^{-1}\mathbf{B}$ criteria, but third by the tr **W** criterion. If we were not interested in further search procedures on our preliminary models, we would have chosen one of these two best partitions as our empirical facies model.

Table 4 is useful in another respect in that it allows us to comment on some of the commonly used classificatory techniques. Judging from the ratings of our preliminary classifications, and recalling how (i.e., by which Procedure) they were obtained, we can empirically conclude that: (1) For a given option of the distance coefficient (i.e., standardization of the data), the unweighted pair-group method tends to give better results than the weighted pair-group method in the situation of Options 1 and 2, but slightly poorer results in the situation of Option 3; (2) item-wise standardization of the data prior to the computation of the distance coefficients gives poorest results by both the weighted and unweighted pair-group methods, and is not to be recommended; (3) because the distance coefficients obtained after item-wise standardization have a simple functional relation to the item-wise (Q-mode) correlation coefficients, the use of these coefficients as a similarity measure is also not recommended. These conclusions should be considered tentative in view of the limited amount of data on which they are based. More empirical assessment of the classificatory techniques would be most useful.

IMPROVEMENT OF PRELIMINARY MODELS

An essential step in our approach is the improvement of each of the preliminary models. This is a search procedure intended to improve the quality of each preliminary classification through the reduction of the tr **W** value. The improvement is accomplished iteratively and through the nearest-neighbor algorithm, described by Demirmen (1969). At the end of the iterations, a new partition, called the limit partition, is reached beyond which no further improvement by the nearest-neighbor algorithm is possible. A measure of the distance between the initial partition and the partition generated during an iteration is given by the number of core items, i.e., items that are not displaced from their original classes in the initial partition (Demirmen, 1969). The value of tr **W** is computed again in the y-space.

The results of the application of the nearest-neighbor algorithm to our preliminary models are tabulated iteration-by-iteration in Table 5. For each preliminary partition, the progressive reduction in tr **W**, marking successive improvements relative to this criterion, is evident. In all the six situations, the most pronounced improvement takes place from the first to the second iteration, the improvement in general becoming progressively less marked as the iterations continue. The preliminary classifications 1-A and 3-B reach their limit situations in 5 iterations, whereas the preliminary classification 3-A reaches its limit situation in 16 iterations. Of the limit partitions, the limit 3-A contains the smallest number (136 out of 200) of core items, indicating the greatest amount of reshuffling during the iterations. The limit 1-A, with 184 core items, indicates the least amount of reshuffling. As expected, the number of core items in general decreases as the iterations proceed, although a few, minor reversals are present.

For interest, the values that Λ and tr $\mathbf{W}^{-1}\mathbf{B}$ assume in the y-space during each iteration also are tabulated in Table 5. It is evident that, with a few exceptions, Λ becomes smaller at successive iterations, implying that improvement by the tr $\mathbf{W}$ criterion is more or less paralleled by an improvement by the Λ criterion. Here again, the tr $\mathbf{W}^{-1}\mathbf{B}$ criterion exhibits a more erratic behavior although, in the situation of the preliminary partitions 1-A and 1-B, it also marks a progressive improvement with iterations.

SELECTION OF BEST IMPROVED MODEL

Having improved our preliminary models, we now proceed to select the best of the improved models. To do this, we again apply our assessment criteria to these partitions. The pertinent data, along with the ratings, are shown in Table 6. The values listed for each criterion are the same as those listed for the limit partitions in Table 5. We note that the poorest ratings correspond to the limit partitions 1-A, 1-B, and to some extent, 3-B. The limit 2-A has intermediate rating, whereas the limit 2-B and limit 3-A rate as the two best. A choice between the limit partitions 2-B and 3-A should probably favor the limit 2-B. This is because the limit 2-B rates better by both the Λ and the tr $\mathbf{W}^{-1}\mathbf{B}$ criteria, rating only as the next best by the tr $\mathbf{W}$ criterion. Furthermore, the values of tr $\mathbf{W}$ for these two limit situations are so close (4.31 and 4.30) that we can ignore as insignificant the higher rating of the limit 3-A by the tr $\mathbf{W}$ criterion. With these considerations, we select the limit partition 2-B as the best of our improved models.

There is, in general, no correspondence in the relative ratings of a preliminary partition and its limit situation. The preliminary 1-A, for example, rated highly among the preliminary classifications, whereas the limit 1-A rates poorly among the limit partitions.

Whether the amount of reshuffling of the items has any bearing on the quality of a particular partition cannot be ascertained, although there is a suggestion that this may be the situation. The limit partitions 1-A and 1-B, with the poorest ratings, both contain a large number of core items and represent relatively little

Table 5. - Improvement of preliminary models through application of iteration shown is iteration during which limit situation deimprovement compared to partition of immediately pre-

	ITERATION						
	1	2	3	4	5	6	7
PREL. 1-A							
tr W	4.80	4.69	4.65	4.63	4.61		
Λx10^9	15.23	10.03	8.86	8.04	7.36		
tr W^{-1}B	50.15	53.04	53.79	54.11	54.70		
N.C.I.	191	188	186	184	184		
PREL. 1-B							
tr W	4.92	4.77	4.71	4.67	4.60	4.57	4.56
Λx10^9	25.23	13.56	11.80	10.97	9.11	7.95	7.90
tr W^{-1}B	47.03	51.37	51.60	51.78	52.42	53.28	53.31
N.C.I.	188	185	182	176	174	173	173
PREL. 2-A							
tr W	5.58	4.89	4.61	4.46	4.39	4.38	
Λx10^9	13.82	5.98	5.42	5.04	4.68	4.72*	
tr W^{-1}B	63.45	60.42*	57.56*	56.25*	55.97*	55.90*	
N.C.I.	177	163	156	151	150	150	
PREL. 2-B							
tr W	5.84	5.16	4.71	4.41	4.32	4.31	
Λx10^9	28.43	9.93	7.06	5.66	4.19	4.12	
tr W^{-1}B	59.78	59.60*	57.13*	54.73*	56.00	55.97*	
N.C.I.	177	161	148	143	141	141	
PREL. 3-A							
tr W	6.04	4.98	4.75	4.61	4.54	4.50	4.48
Λx10^9	123.6	11.78	9.10	7.25	6.52	6.73*	6.52
tr W^{-1}B	49.44	56.30	55.52*	55.56	55.35*	54.65*	54.71
N.C.I.	165	154	147	142	141	140	140
PREL. 3-B							
tr W	5.96	4.98	4.77	4.67	4.62		
Λx10^9	92.98	9.36	7.90	7.31	6.33		
tr W^{-1}B	49.53	57.14	55.74*	54.84*	55.27		
N.C.I.	170	161	155	152	152		

reshuffling from the preliminary partitions. For the limit partitions 2-B and 3-A, with the highest ratings, the situation is reversed. This condition seems to suggest an inverse relation between the amount of reshuffling and the quality of a limit partition. The limit partition 3-B, however, with a poor rating and a relatively small number of core items, sheds doubt on this supposed relation. The matter seems interesting enough to warrant further nearest-neighbor algorithm. For each preliminary classification, last was reached. N.C.I. - number of core items (out of 200). Sign "*" denotes ceding iteration.

	ITERATION								
	8	9	10	11	12	13	14	15	16
PREL. 3-A									
tr $\underline{W}$	4.46	4.44	4.38	4.37	4.32	4.31	4.31	4.31	4.30
$\Lambda x 10^{-9}$	6.37	6.32	5.33	5.38*	4.63	4.47	4.43	4.42	4.31
tr $\underline{W}^{-1}\underline{B}$	54.64*	54.23*	54.89	54.69*	55.42	55.70	55.73	55.72*	55.78
N.C.I.	139	140	139	136	137	136	137	136	136

investigation.

FURTHER IMPROVEMENT BY REARRANGEMENT

Whereas the limit partition 2-B is the best of our improved classifications, almost certainly it is not the absolute best of all the possible classifications that represent the partition of 200 items into 23 classes. We would expect other but unknown partitions with 23 classes to rate better than the limit 2-B. It may be worthwhile, therefore, to attempt to further improve the limit 2-B. At this point, however, we run into a methodological problem. We know of no algorithm that systematically further improves upon our limit partitions. Trial by random partitions is one possible method, but would be forbiddingly expensive. Instead, we make a trial based on substantive judgment.

Our approach is as follows: We inspect the best of our improved partitions (limit 2-B) for all items that, on substantive (i.e., geological) grounds, seem to be out of place. We reassign these items to what we consider to be their natural classes. This reassignment is effected again on substantive grounds, incorporating all that we know about the geological properties of the samples. We then apply the nearest-neighbor algorithm to the rearranged partition. As before, this algorithm is applied in the y-space. If the limit partition rates better than the existing best (limit 2-B), we accept the new limit partition; otherwise we reject it (we try this only once). Because the rearrangement purports to remove inconsistencies from the existing best, it is not unreasonable to expect that the limit situation of the rearranged partition will rate better than the existing best. The recognition of the items that seem to be out of place in the existing best classification

Table 6. - Assessment of limit situations of six preliminary classifications. N.C.I. - number of core items (out of 200); N.I. - number of iterations performed to reach limit situation. For given criterion, rating 1 denotes best, rating 2 denotes second best, etc.

	LIMIT					
	1-A	1-B	2-A	2-B	3-A	3-B
CRITERION						
tr $\underline{W}$	4.61	4.56	4.38	4.31	4.30	4.62
$\Lambda x10^9$	7.36	7.90	4.72	4.12	4.31	6.33
tr $\underline{W}^{-1}\underline{B}$	54.70	53.31	55.90	55.97	55.78	55.27
RATING						
tr $\underline{W}$	5	4	3	2	1	6
Λ	5	6	3	1	2	4
tr $\underline{W}^{-1}\underline{B}$	5	6	2	1	3	4
N.C.I.	184	173	150	141	136	152
N.I.	5	7	6	6	16	5

and their reassignment to their natural classes is facilitated by the use of the *z*-scores. These scores have a high substantive value and help compare the geologic properties of the items and classes.

Upon the application of the nearest-neighbor algorithm, the rearranged form of the limit 2-B, after some minor reshuffling, reached its own limit in 3 iterations. The values of our criteria pertaining to the new limit partition are: tr $\mathbf{W}$ = 4.28, Λ = 3.62×10^{-9}, and tr $\mathbf{W}^{-1}\mathbf{B}$ = 57.03. If we compare these values with those for the limit 2-B (Table 6), we note that the new limit rates better than the limit 2-B by all three criteria. Hence, we accept the new limit partition . The distribution of the items in classes according to this partition is shown in Table 7.

REDUCTION OF NUMBER OF CLASSES

In principle, the partition shown in Table 7 represents the facies model of our sample space. Each of the classes contained in this partition is capable of reasonable geologic interpretation, as judged by their mean *z*-scores, and can be construed to portray a particular environment of deposition. There is one nagging

Table 7. - Facies partition with 23 classes after rearrangement and improvement of limit 2-B.

Cl. 1	A-71	F-69-2	G-74	B-70	Cl. 16	B-65-2
A-60	B-71	G-69-1	M-75-2	G-70	C-69-1	E-75
C-60	H-71	G-69-2	Cl. 7	H-70	Cl. 17	G-75-1
D-60	I-71	H-69-1	Y-61-3	K-70	Y-75-2	G-75-2
E-60-1	K-71-2	H-69-2	B-61	L-70	A-75	Cl. 21
E-60-2	L-71	I-69-2	F-61	Y-71-1	F-75-1	A-62
F-60	M-71-3	K-69-2	H-61	C-71	F-75-2	F-62-2
G-60	B-75	K-69-3	I-61	F-71	G-75-3	I-62
H-60	J-75	L-69-1	L-61	C-75-1	H-75	D-63
I-60-2	L-75-2	L-69-2	B-62	D-75	Cl. 18	E-63
J-60-1	Cl. 4	M-69-2	E-62	F-75-3	L-62	H-63
J-60-2	B-60-2	M-69-3	H-62	I-75	Y-63-2	I-63
K-60	Y-61-1	E-74-2	J-62	Cl. 10	F-63	I-74-4
L-60-1	C-61	I-74-3	K-62	M-71-1	G-63	J-74-1
L-60-2	D-61	K-74-2	Cl. 8	M-71-2	Y-64-1	M-75-1
M-60	E-61	L-74	D-62	F-74-1	Y-65	Cl. 22
Cl. 2	G-61	M-74-2	F-62-1	F-74-2	B-65-1	Y-69-1
Y-70-1	J-61	Cl. 6	L-63	F-75-4	E-65	D-69-1
Y-70-2	K-61	C-63	D-65	Cl. 11	H-65-2	K-69-1
E-70	M-61	B-64	H-65-1	C-75-2	J-65	Y-74-1
D-71	C-62	F-64	L-65	Cl. 12	M-65	Y-74-2
E-71	Y-63-1	G-64	I-69-1	A-74-0	M-75-3	A-74-3
Cl. 3	K-63	J-64	J-69-2	A-74-1	Cl. 19	B-74-1
Y-60	E-64	A-65	M-69-1	Cl. 13	C-74	H-74
B-60-1	Cl. 5	C-65	F-70	A-61	D-74	I-74-1
I-60-1	Y-69-2	F-65	J-71	Cl. 14	E-74-1	Cl. 23
Y-62-2	Y-69-3	G-65-1	B-74-2	M-63	Cl. 20	I-74-5
A-70	A-69-1	G-65-2	I-74-2	I-65	Y-62-1	J-74-2
C-70	A-69-2	K-65	K-74-1	Cl. 15	G-62	Y-75-1
D-70	B-69-2	K-69-4	M-74-1	B-69-1	A-63	K-75
I-70	C-69-2	K-71-1	Y-75-3	F-69-1	B-63	L-75-1
J-70	D-69-2	A-74-2	Cl. 9	J-69-1	J-63	

problem with this model, however: the large number of classes, which is 23. This many number of classes or facies is impractical to work with in environmental reconstruction. As the next and final step, therefore, the number of classes was reduced to a more practical number, 13. The procedure used to reduce the num-

Table 8. - Final facies partition with 13 classes after reduction of number of classes.

Cl. 1	M-71-1	Cl. 4	I-69-2	M-75-2	D-62	I-69-1
A-60	F-74-2	B-60-2	J-69-1	Cl. 7	F-62-1	J-69-2
C-60	C-75-1	Y-61-1	K-69-2	Y-61-3	G-62	K-69-1
D-60	D-75	C-61	K-69-3	B-61	A-63	M-69-1
E-60-1	F-75-3	D-61	L-69-1	F-61	B-63	Y-74-1
E-60-2	F-75-4	E-61	L-69-2	H-61	J-63	Y-74-2
F-60	I-75	G-61	M-69-2	I-61	B-65-2	B-74-1
G-60	Cl. 3	J-61	M-69-3	L-61	D-65	C-74
H-60	Y-60	K-61	B-74-2	B-62	A-69-2	D-74
I-60-2	B-60-1	M-61	E-74-2	E-62	A-75	E-74-1
J-60-1	I-60-1	C-62	I-74-3	H-62	E-75	H-74
J-60-2	Y-62-2	Y-63-1	K-74-2	J-62	F-75-1	I-74-1
K-60	L-65	K-63	L-74	K-62	F-75-2	K-74-1
L-60-1	A-70	M-63	M-74-2	Cl. 8	G-75-1	M-74-1
L-60-2	C-70	Y-64-1	Cl. 6	C-75-2	G-75-2	M-75-3
M-60	D-70	E-64	C-63	Cl. 9	G-75-3	Cl. 13
Cl. 2	I-70	I-65	B-64	A-74-0	H-75	I-74-5
Y-70-1	A-71	C-69-1	F-64	A-74-1	Cl. 12	J-74-2
Y-70-2	B-71	D-69-2	G-64	Cl. 10	L-62	Y-75-1
B-70	H-71	H-69-2	J-64	A-61	Y-63-2	K-75
E-70	I-71	Cl. 5	A-65	A-62	F-63	L-75-1
F-70	J-71	Y-69-2	C-65	F-62-2	G-63	
G-70	K-71-2	Y-69-3	F-65	I-62	L-63	
H-70	L-71	A-69-1	G-65-1	D-63	Y-65	
J-70	M-71-3	B-69-1	G-65-2	E-63	B-65-1	
K-70	A-74-3	B-69-2	K-65	H-63	E-65	
L-70	I-74-2	C-69-2	K-69-4	I-63	H-65-1	
Y-71-1	Y-75-2	F-69-1	K-71-1	I-74-4	H-65-2	
C-71	Y-75-3	F-69-2	M-71-2	J-74-1	J-65	
D-71	B-75	G-69-1	A-74-2	M-75-1	M-65	
E-71	J-75	G-69-2	F-74-1	Cl. 11	Y-69-1	
F-71	L-75-2	H-69-1	G-74	Y-62-1	D-69-1	

ber of classes will not be given here for lack of space. The guiding principle adopted in reducing the number of classes was that the technique utilized should be as objective as possible, but that the results obtained should be capable of reasonable geologic interpretation, as judged from the mean z-scores of the classes. Thus a middle course was used between what may be regarded as purely objective and purely subjective methods. The ad hoc procedure devised by the writer for this purpose allows stepwise reduction of the number of classes to an arbitrary level and also permits an inspection and geologic assessment of the results at each step.

For this situation, it was judged from the geologic properties of the classes defined, that a partition with 13 classes would provide the optimum facies model for environmental reconstruction, and that further reduction in the number of classes would result in the loss of valuable geologic information. The partition with 13 classes was accepted therefore as the revised and final facies model of our sample space. This partition is shown in Table 8.

A relevant question comes to mind at this juncture. If it could be decided beforehand that 23 classes are impractical to work with, why did we not partition our sample space to a smaller number of classes, say 13, in the first place? The answer to this question can be found in the guidelines adopted in defining the preliminary classes in the dendrograms (see, Definition of Preliminary Models). In particular, the principle that the classifi-

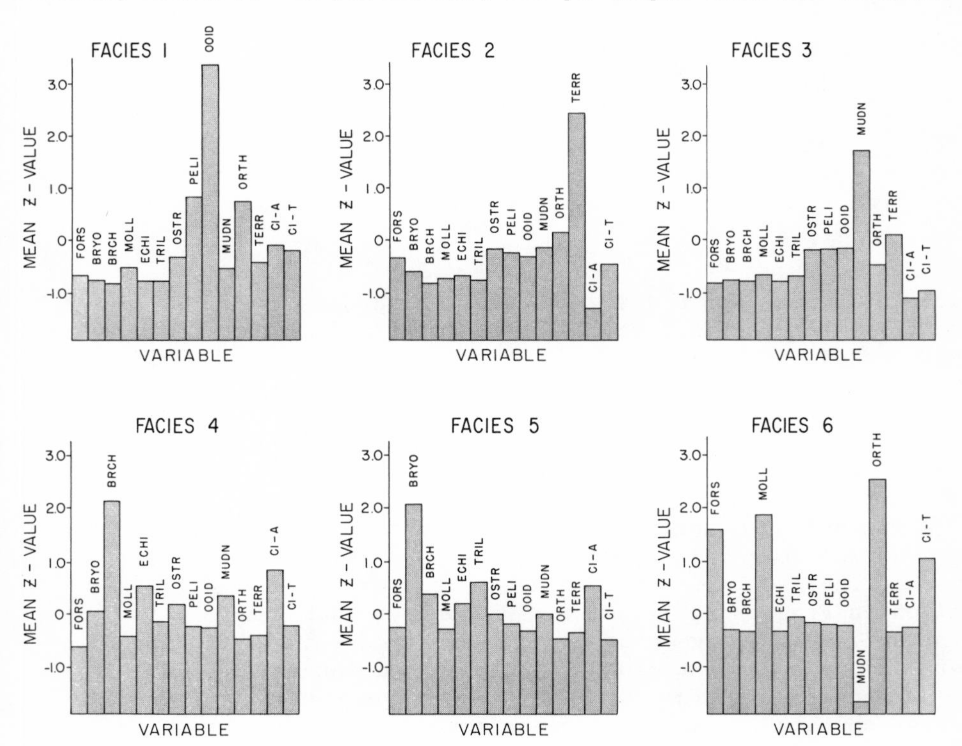

Figure 4. - Mean z-scores for Facies 1 through 6.

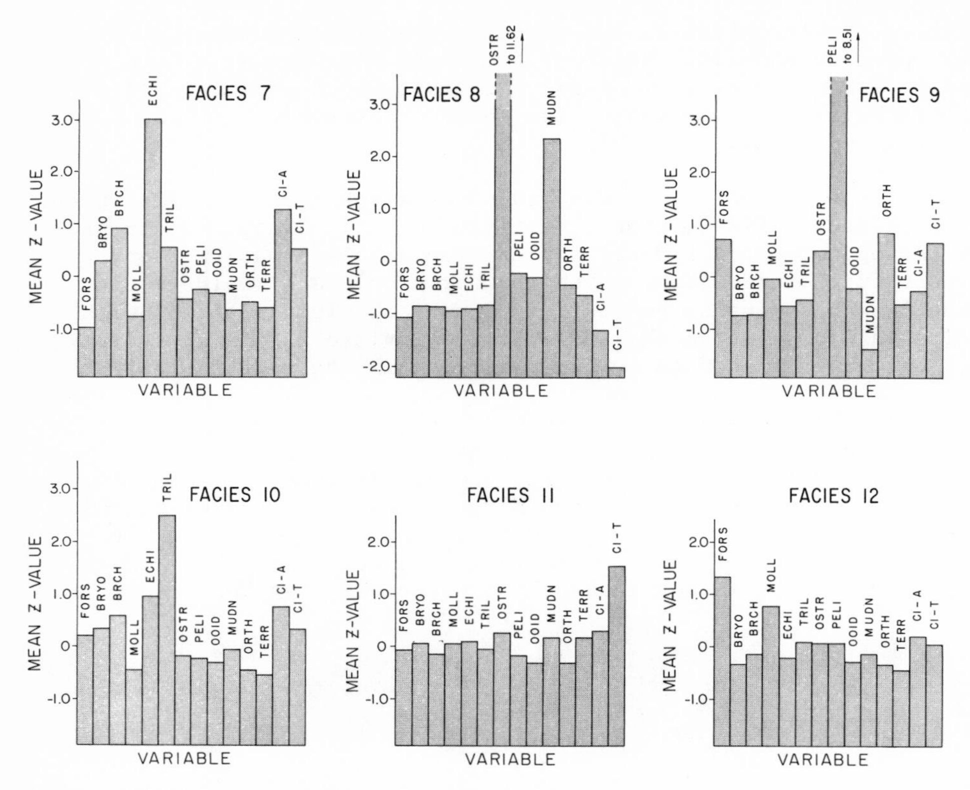

Figure 5. - Mean z-scores for Facies 7 through 12.

cations obtained from the six dendrograms should be similar to each other, and the constraint that the classifications so defined should contain an equal number of classes, necessitated working with a large number of classes at the outset.

FACIES ATTRIBUTES

A convenient description of our facies is provided by the mean z-scores computed for each of the 13 classes in our model. A graphic representation of the mean z-scores for the 13 classes is given in Figures 4, 5, and 6. We observe that Facies 1 is characterized by a high content of OOID (ooids, $\bar{z}$ = 3.35) and a moderately high content of PELI (pellets and intraclasts, $\bar{z}$ = 0.83). Facies 2 is high in TERR (terrigenous material), with $\bar{z}$ = 2.42 on this variable. Another noteworthy attribute of Facies 2, however, is its relatively low mean z-score on CI-A (coarseness index for allochems, $\bar{z}$ = -1.30). Hence Facies 2 is characterized by the relative abundance of terrigenous material on the one hand,

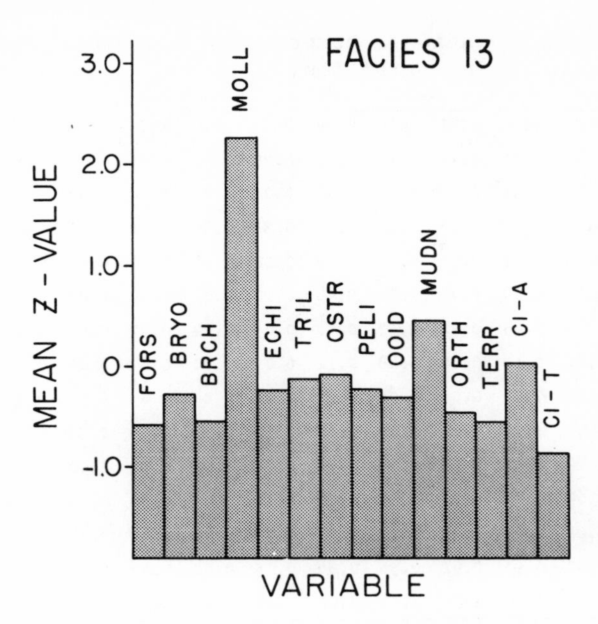

Figure 6. - Mean z-scores for Facies 13.

and the relative fineness of the allochemic constituents on the other hand. (We assume that the CI-A in a rock is monotonically related to the average particle size of allochems in that rock. A similar assumption applies in the interpretation of CI-T.) With similar arguments the salient characteristics of the remaining facies become evident from an inspection of Figures 4, 5, and 6.

WITHIN-FACIES VARIATION

An assessment of the within-facies variation in attributes can be obtained from the standard deviations and absolute values of the coefficients of variation in z-scores computed for each of the 13 classes in our model. In general, if we let n_h be the number of items in the h-th class $(\Sigma n_h = n)$, z_{hki} the z-score of the k-th item, contained in the h-th class, relative to the i-th variable, and $z_{h.i}$ the mean z-score for the h-th class relative to the i-th variable, we can compute some $S(m\text{x}p)$ and $C(m\text{x}p)$ matrices such that

$$S = (s_{h.i}),\ s_{h.i} = \sqrt{\frac{\sum_{k=1}^{n_h} (z_{hki} - z_{h.i})^2}{n_h - 1}}\ ,\quad C = (|c_{h.i}|),\ |c_{h.i}| = \left|\frac{s_{h.i}}{z_{h.i}}\right| \quad k\varepsilon h;\ \begin{array}{l} h=1,\ldots,m; \\ i=1,\ldots,p; \end{array} \qquad (12)$$

Table 9. - Standard deviations, computed in z-scores, over classes in final facies partition.

	FORS	BRYO	BRCH	MOLL	ECHI	TRIL	OSTR	PELI	OOID	MUDN	ORTH	TERR	CI-A	CI-T
CLASS 1	0.27	0.08	0.13	0.35	0.17	0.15	0.21	0.99	0.89	0.75	1.19	0.51	0.47	0.31
CLASS 2	0.68	0.34	0.13	0.41	0.12	0.32	0.42	0.01	0.0	0.78	0.93	0.63	0.40	0.58
CLASS 3	0.41	0.28	0.28	0.42	0.17	0.34	0.53	0.31	0.40	0.48	0.00	0.64	0.62	0.62
CLASS 4	0.45	0.80	0.54	0.51	0.75	0.72	0.78	0.00	0.18	0.50	0.02	0.31	1.35	0.58
CLASS 5	0.47	0.94	0.71	0.29	0.44	1.05	0.48	0.10	0.0	0.37	0.00	0.25	0.53	0.42
CLASS 6	0.65	0.37	0.49	0.63	0.22	0.53	0.41	0.05	0.24	0.07	0.44	0.60	0.42	0.79
CLASS 7	0.17	0.24	0.68	0.18	0.66	0.68	0.12	0.01	0.00	0.44	0.00	0.16	0.67	0.94
CLASS 8	0.0	0.0	0.0	0.0	0.0	0.0	0.0	0.0	0.0	0.0	0.0	0.0	0.0	0.0
CLASS 9	0.86	0.03	0.04	1.02	0.24	0.0	1.36	2.64	0.13	0.08	0.53	0.23	0.61	0.46
CLASS 10	0.60	0.46	0.74	0.32	0.81	1.18	0.40	0.02	0.00	0.53	0.00	0.20	0.57	1.10
CLASS 11	0.47	0.48	0.64	0.74	0.71	0.63	0.78	0.11	0.0	0.57	0.39	0.74	0.72	0.88
CLASS 12	0.68	0.38	0.62	0.66	0.47	0.63	0.60	0.71	0.02	0.62	0.30	0.26	0.52	0.83
CLASS 13	0.74	0.54	0.32	0.97	0.41	0.56	0.66	0.05	0.00	0.43	0.01	0.11	0.19	0.09
GRAND	1.00	1.00	1.00	1.00	1.00	1.00	1.00	1.00	1.00	1.00	1.00	1.00	1.00	1.00

where $s_{h,i}$ and $c_{h,i}$ are the standard deviation and the coefficient of variation, respectively, of the i-th variable over the h-th class. In our example m = 13 and p = 14. The **S** and **C** matrices, both of order (13 x 14), for our model are shown in Tables 9 and 10, respectively.

Table 10. - Coefficients of variation (absolute values), computed in z-scores, over classes in final facies partition.

	FORS	BRYO	BRCH	MOLL	ECHI	TRIL	OSTR	PELI	OOID	MUDN	ORTH	TERR	CI-A	CI-T
CLASS 1	0.41	0.11	0.16	0.69	0.22	0.19	0.67	1.19	0.26	1.41	1.65	1.22	4.62	1.58
CLASS 2	2.10	0.58	0.16	0.57	0.17	0.43	2.50	0.04	0.0	5.49	7.48	0.26	0.31	1.29
CLASS 3	0.51	0.37	0.36	0.64	0.21	0.50	2.88	1.85	2.48	0.28	0.00	6.15	0.57	0.66
CLASS 4	0.72	20.32	0.26	1.22	1.44	5.10	4.05	0.00	0.69	1.45	0.05	0.78	1.62	2.69
CLASS 5	1.82	0.45	1.81	1.00	2.12	1.72	249.95	0.51	0.0	134.91	0.00	0.72	1.01	0.85
CLASS 6	0.41	1.25	1.53	0.33	0.71	7.75	2.47	0.26	1.11	0.04	0.17	1.80	1.68	0.74
CLASS 7	0.18	0.80	0.75	0.23	0.22	1.19	0.28	0.06	0.00	0.69	0.00	0.27	0.52	1.81
CLASS 8	0.0	0.0	0.0	0.0	0.0	0.0	0.0	0.0	0.0	0.0	0.0	0.0	0.0	0.0
CLASS 9	1.23	0.05	0.05	45.18	0.43	0.0	2.69	0.31	0.62	0.06	0.63	0.44	2.60	0.69
CLASS 10	2.95	1.48	1.25	0.69	0.84	0.47	2.10	0.08	0.00	7.72	0.00	0.37	0.75	3.50
CLASS 11	6.19	7.82	4.30	12.72	7.29	11.16	2.88	0.64	0.0	3.27	1.27	4.24	2.39	0.58
CLASS 12	0.52	1.18	4.19	0.86	2.25	7.26	9.79	11.82	0.05	4.29	0.86	0.56	2.75	23.95
CLASS 13	1.26	1.93	0.58	0.42	1.69	4.54	9.36	0.22	0.00	0.94	0.02	0.20	9.73	0.10

It is evident from these tables that the greatest absolute variation (standard deviation) occurs with the variable PELI in Class 9, next followed by OSTR in the same class, and CI-A in Class 4. Relative to the mean, however, the greatest variation (coefficient of variation) occurs with the variables OSTR and MUDN in Class 5.

SUBFACIES

An inspection of the *z*-scores in our facies model reveals that some of the items in certain classes have distinct geologic properties that are not shared by all the items belonging to these classes. The mean *z*-values, being measured over all the items in a given class, do not bring out the peculiarities of all the items in a class and mask those attributes that are shared by relatively few items in that class. It is, of course, impractical to consider all the items with such peculiarities in our classification. It is feasible, however, to single out those items that, in our judgment, merit separate recognition. In this manner we can divide certain facies into subfacies and obtain a more detailed

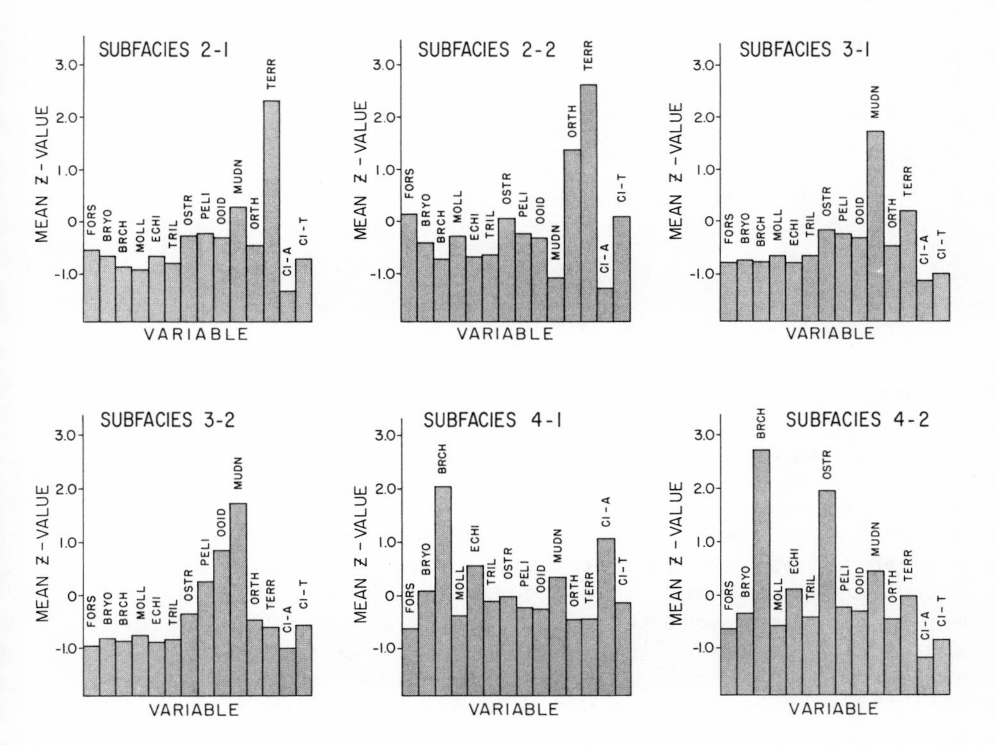

Figure 7. - Mean *z*-scores for Subfacies 2-1, 2-2, 3-1, 3-2, 4-1, and 4-2.

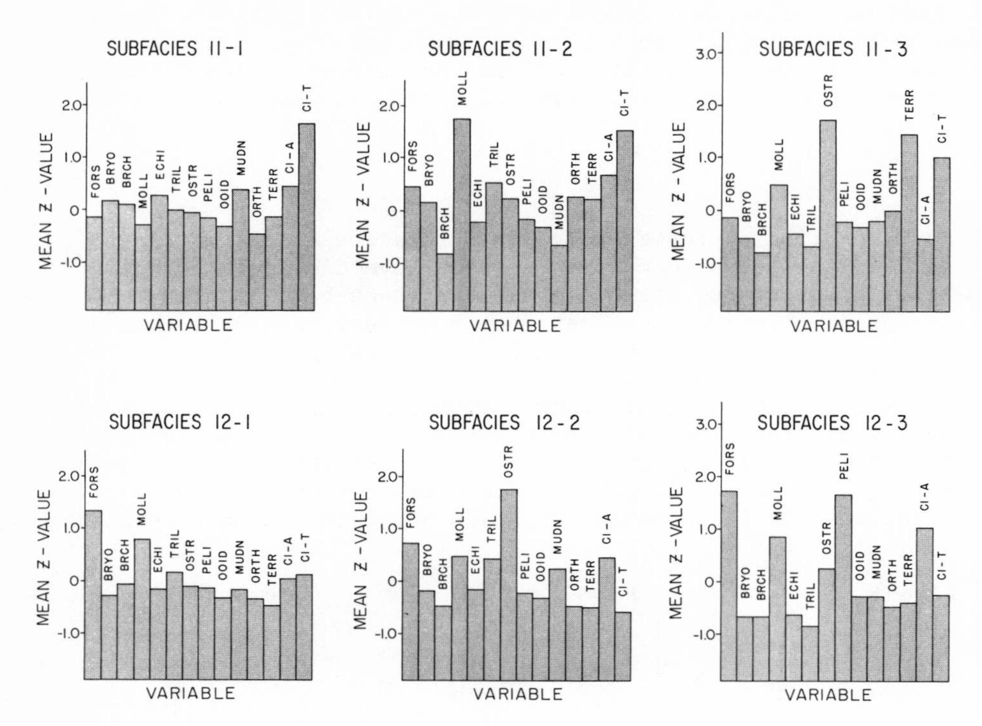

Figure 8. - Mean z-scores for Subfacies 11-1, 11-2, 11-3, 12-1, 12-2, and 12-3.

or refined partition. For a given facies, one of the subfacies defined in this manner will, of necessity, contain those left-over items that do not possess special properties. Such subfacies may be termed neutral subfacies.

The refinement of our facies partition with the principle noted above involved the division of Facies 2, 3, and 4 into two subfacies each, and the division of Facies 11 and 12 into three subfacies each. Thus, altogether 12 subfacies were recognized. For a given facies, the neutral subfacies is designated as the first facies (e.g., Subfacies 2-1 for Facies 2, Subfacies 3-1 for Facies 3). The mean z-scores corresponding to the 12 subfacies are graphically portrayed in Figures 7 and 8. A comparison of these figures with Figures 4 and 5 reveals in what manner, if any, a particular subfacies stands out relative to its parent facies. It is evident, for example, that relative to the parent Facies 2, Subfacies 2-1 is a neutral subfacies, whereas Subfacies 2-2 brings together those items of Facies 2 that are relatively high in ORTH (orthospar) content. Both of these subfacies, however, have high TERR and low CI-A values, properties that also characterize the parent Facies 2.

A discussion of the environmental interpretation of our facies and subfacies lies outside the scope of the present contribution and will be given in a future paper.

CONCLUSION

In a facies study of the Honaker Trail limestones, mathematical search procedures were used to arrive at more meaningful empirical facies models. These search procedures involve essentially the use of competitive classificatory schemes, assessment, by intuitively appealing objective criteria, of the resulting facies partitions, and further improvement of these partitions. The use of such search procedures could be recommended in other studies of facies modeling or in other classification problems in which the investigator is faced with a variety of techniques to choose from with no reliable a priori knowledge on the relative mertis of these techniques.

ACKNOWLEDGMENTS

The writer is indebted to Professors John W. Harbaugh and Paul Switzer, both of Stanford University, for their counsel during the course of the work and to Dr. John C. Davis of the Kansas Geological Survey for reading the manuscript. Financial support for the work was provided by a NATO Science Fellowship to the writer and by a National Science Foundation grant (NSF GP 4514) to Professor Harbaugh. The School of Earth Sciences at Stanford University furnished generous computer time for the necessary calculations.

REFERENCES

Anderson, A. J. B., 1971, Numeric examination of multivariate soil samples: Jour. Intern. Assoc. Math. Geol., v. 3, no. 1, p. 1-14.

Chayes, F., 1962, Numerical correlation and petrographic variation: Jour. Geology, v. 70, no. 4, p. 440-452.

Demirmen, F., 1969, Multivariate procedures and FORTRAN IV program for evaluation and improvement of classifications: Kansas Geol. Survey Computer Contr. 31, 51 p.

Demirmen. F., 1971, Counting error in petrographic point-count analysis: A theoretical and experimental study: Jour. Intern. Assoc. Math. Geol., v. 3, no. 1, p. 15-41.

Feldhausen, P. H., 1970, Ordination of sediments from the Cape Hatteras continental margin: Jour. Intern. Assoc. Math. Geol., v. 2, no. 2, p. 113-129.

Herdan, G., 1953, Small particle statistics: Elsevier Publ. Co., New York, 520 p.

Merriam, D. F., 1970, Comparison of British and American Carboniferous cyclic rock sequences: Jour. Intern. Assoc. Math. Geol., v. 2, no. 3, p. 241-264.

Miller, R. L., and Kahn, J. S., 1962, Statistical analysis in the geological sciences: John Wiley & Sons, Inc., New York, 483 p.

Sokal, R. R., and Sneath, P. H. A., 1963, Principles of numerical taxonomy: W. H. Freeman and Co., San Francisco, 359 p.

Switzer, P., 1970, Numerical classification, in Geostatistics, a colloquium: Plenum Press, New York, p. 31-43.

MONTE CARLO SIMULATION OF SOME FLYSCH DEPOSITS FROM THE EAST CARPATHIANS

Mircea Dumitriu and Cristina Dumitriu

Romanian Geological Institute

ABSTRACT

The mathematical simulation of a series of Cretaceous-Paleogene flysch deposits from the East Carpathians (Romania) on an IBM 360/40 computer was made to obtain information regarding the sedimentation mechanism of flysch deposits. Thickness of strata for two states (sandstone and shale) of different age and geographic location was measured. The mathematical simulation of the flysch sequences was accomplished by a random extraction of shale thickness from a lognormal distribution and sandstone thickness from an empirical distribution. A comparison between the simulated and real curves, representing the variation of sandstone contents, established by the computation and automatic graphic representation of sandstone percentages in fix vertical intervals, revealed similarity of two series.

INTRODUCTION

Geological simulation has recorded a remarkable rise in the past few years. This is especially because of the imitation of geological processes with mathematical models using high-speed electronic computers (see Harbaugh and Merriam, 1968; Krumbein and Graybill, 1965).

By systematically changing the process elements of the conceptual process-response model of a geological phenomenon, various models of the response elements can be obtained. Similarity between the simulated and the real response-model shows that the

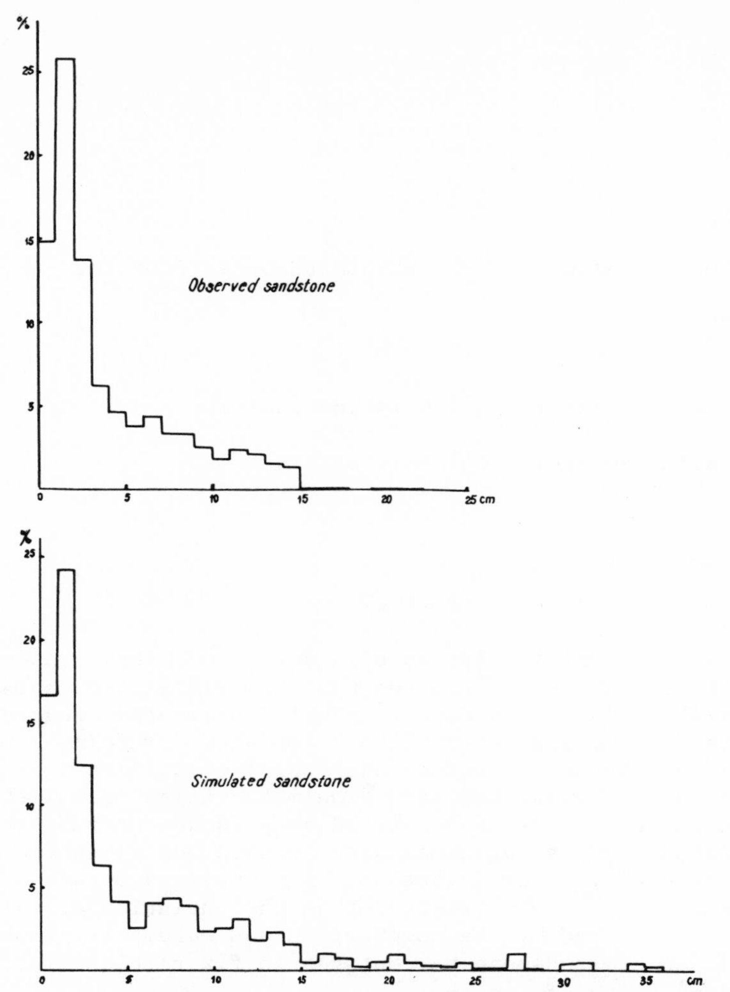

Figure 1. - Frequency-distribution histograms of observed and simulated sandstone thickness (Oligocene sequence, Vinetisu).

simulated process-model describes in an objective manner the real process model.

The natural process can be described by deterministic variables, if the state of the system at a given moment in time or space is completely predetermined, or by stochastic variables if the behavior of the system is changing in a random manner. Many geological processes, for example sedimentary process, contain some random element and can be described statistically. It is known that many geological frequency distributions belong to familiar

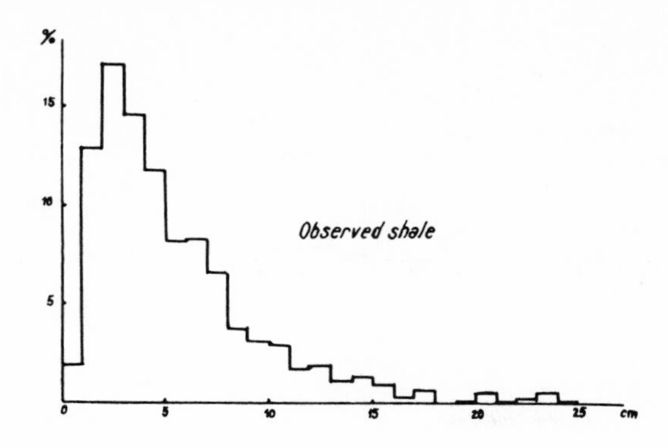

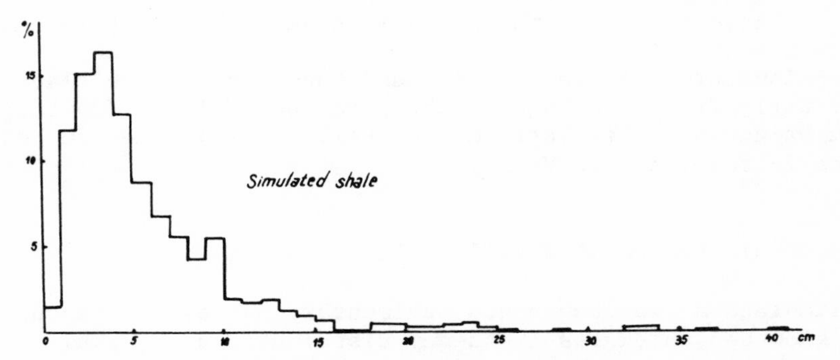

Figure 2. - Frequency-distribution histograms of observed and simulated shale thickness (Oligocene sequence, Vinetisu).

models (gaussian, lognormal, gamma, exponential, etc.). We have used for the study in behavior of geological systems in time or space therefore, Monte Carlo simulation methods. These consist of the generation of pseudorandom numbers for the use of "drawing" random samples from known frequency distributions.

This paper contains the results of the application of Monte Carlo simulation methods to generating synthetic flysch columns (binary sandstone-shale systems). The object is to establish the predictive possibilities for the development of sedimentary formations in zones of economic interest.

The primary data that we applied in this simulation are represented by bed thickness measurements. The thickness measurements were made in Cretaceous-Paleogene flysch series of the East Carpath-

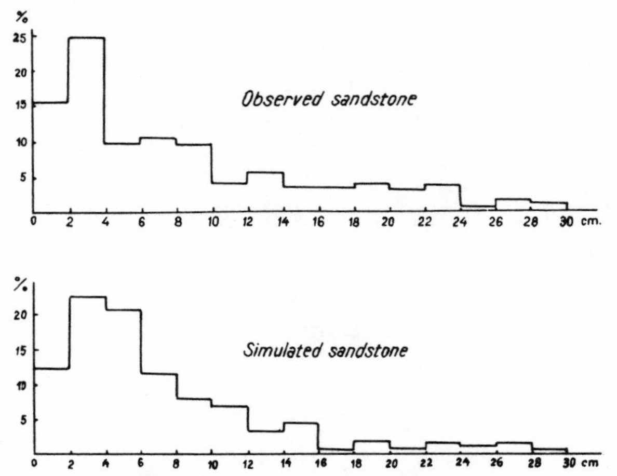

Figure 3. - Frequency-distribution histograms of observed and simulated sandstone thickness (Eocene sequence, Damacusa).

ians. Sections are (1) glauconitic sandstone zone of the black shales of Early Cretaceous age in the Sbrancani Valley; (2) Eocene beds from Damacusa Valley-Vatra Moldovitei, and (3) Vinetisu beds (Oligocene)in the Vinetisu Valley.

SIMULATION OF STRATIGRAPHIC SEQUENCES

To simulate a two-components sedimentary series, it is sufficient to know bed-thickness frequency distributions. For all sedimentary series with which mathematical simulation trials were

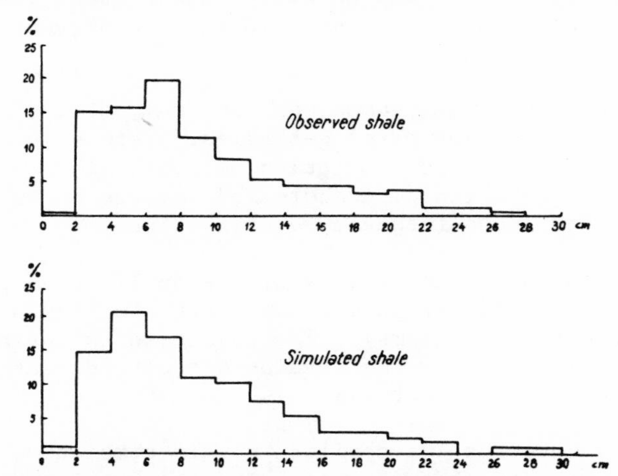

Figure 4. - Frequency-distribution histograms of observed and simulated shale thickness (Eocene sequence, Damacusa).

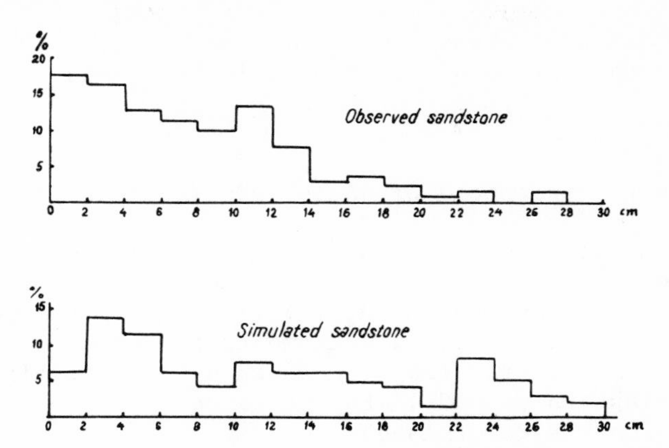

Figure 5. - Frequency-distribution histograms of observed and simulated sandstone thickness (Lower Cretaceous sequence, Sbrancani).

performed, histograms were drawn for the frequency distributions of sandstone and shale (Fig. 1, 2, 3, 4, 5, and 6). Thickness of shale units for all series were lognormal. Thickness of sandstone units also were lognormal. Thickness of sandstone units from the Damacusa Series also were lognormal. Thickness of sandstone units from the Vinetisu and Sbrancani Series, however, were considered to be from an empirical distribution. We observed a strong left asymmetry of the sandstone distribution compared to the shale distribution. In the first stage of our research, simulation was achieved by a random selection of thickness from a lognormal distribution for shales and sandstones of the Damacusa Series and from an empirical distribution for sandstones from the Vinetisu and Sbrancani Series.

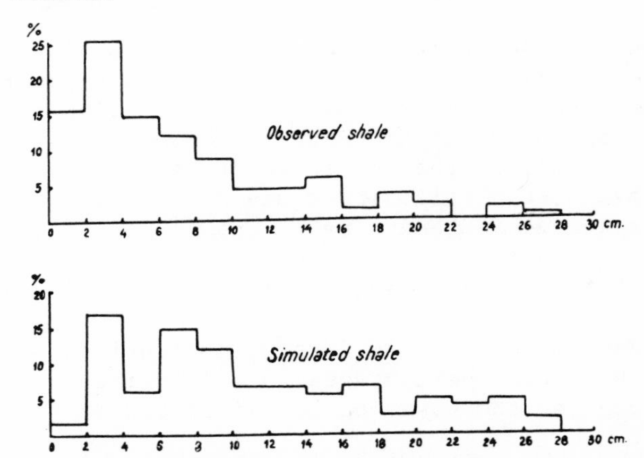

Figure 6. - Frequency-distribution histograms of observed and simulated shale thickness (Lower Cretaceous sequence, Sbrancani).

For thickness distribution of shales and sandstones of the Damacsa Series, the following transformation was used

$$L = ln\ X\ ,$$

where X is the thickness of shale beds (in meters).

A FORTRAN IV program, written by us, contains read statements for bed thickness (measured in different sections), selection statements for shale and sandstone, logarithmatic statements for shale thickness, and computation statements for mean and standard deviation of normal distributions obtained by transforming the lognormal distribution of beds. For sandstone thickness, the frequency function was determined.

A second part of the program is an adaptation of a program written by Harbaugh and Bonham-Carter (1970) and includes: statements for mathematical simulation of sedimentary series for which there are computed means and standard deviations (the example of shales and sandstones of the Damacusa, having a lognormal distribution, and cumulative frequency distributions (the example of sandstones of the Vinetisu and Sbrancani) having an empirical distribution.

Three subroutines for generating pseudorandom numbers: RANDU (from System IBM/360 scientific subroutines package, version III), LGAUSS and RANEMP (from Harbaugh and Bonham-Carter, 1970) were used for the random selection of bed thickness during the simulation run. The subroutine RANDU generates uniformly distributed numbers, the subroutine LGAUSS, lognormal distributed numbers, and the subroutine RANEMP, empirical distributed numbers.

Simulation of a sedimentary series was accomplished in the following stages: (1) selection by means of RANDU-generator of an initial state (sandstone or shale); (2) selection from the sandstone distribution, by means of RANEMP, or from the shale distribution, by means of LGAUSS, the initial state thickness; and (3) automatic passage to the following state of the system, the thickness of the respective bed being determined in the same manner as at point 2, until the thickness of the simulated series equalled the thickness of the observed series (Table 1).

A third part of the FORTRAN IV program provides statements for computation of sandstone percentages for fixed intervals with regard to time both for the observed and simulated series (Fig. 7).

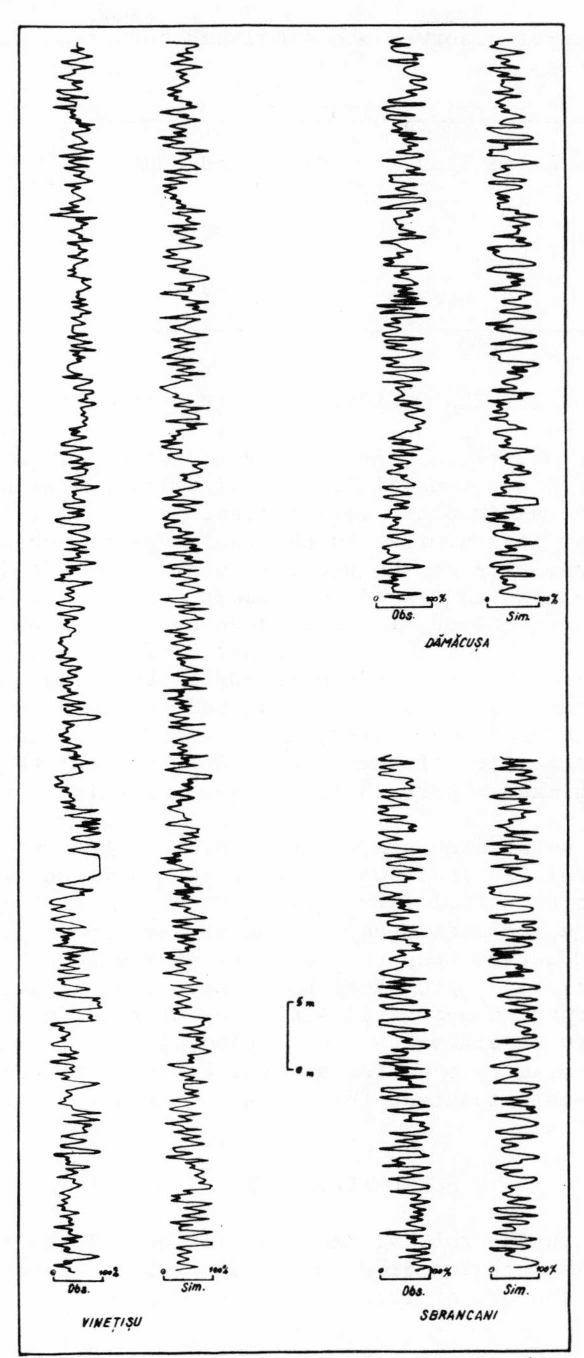

Figure 7. - Sandstone-percentage curve for some observed and simulated flysch series from East Carpathians.

Table 1. - Number of observed and simulated beds from studied sequences.

Sequence	Observed beds	Simulated beds	$\frac{\text{Sim. beds}}{\text{Obs. beds}}$ 100
Vinetisu	1864	1799	-3.5
Damacusa	528	538	1.9
Sbrancani	286	299	4.5

COMPARISON BETWEEN OBSERVED AND SIMULATED SERIES

In order to understand sedimentary phenomena, it is necessary to determine the process-model (the total process-elements) responsible for the sedimentary series (response-model). Simulated sequences more or less similar to the real ones are obtained by changing the parameters of the process-model. In this situation the process-model, which served as a basis for generating the random stratigraphic sequence, is considered the real process-model of the sedimentary phenomenon. It is not difficult to observe, from the comparison between bed-thickness distributions of the simulated sandstones and shales and the bed-thickness distributions of the real sandstones and shales, the similarity which exists between them. Application of the χ^2-test demonstrated that both series of bed thickness pertain to the same distribution.

Comparison between curves, which represent the variation of observed and simulated sandstone percentages on fixed vertical intervals, shows the similarity between the two series (Fig. 7). Distances between the maxima and minima of the sandstone contents of the simulated curves keep the same values recorded by the real curve. The theoretical procedure has a practical importance because it represents an automatic simulator of stratigraphic sequences necessary to elaborate exploration plans for ore deposits, using only the parameters of the bed-thickness distributions, which seem to remain constant for the same formation.

ACKNOWLEDGMENTS

We wish to thank Prof. V. Ianovici for help in pursuing our research. We also are indebted to the Geological Institute (Bucharest) for support of this work.

REFERENCES

Harbaugh, J. W., and Merriam, D. F., 1968, Computer applications in stratigraphic analysis: John Wiley & Sons, New York, 282 p.

Harbaugh, J. W., and Bonham-Carter, G. F., 1970, Computer simulation in geology: John Wiley & Sons, New York, 575 p.

Krumbein, W. C., and Graybill, F. A., 1965, An introduction to statistical models in geology: McGraw-Hill Book Co., New York, 475 p.

DIFFUSION MODEL OF SEDIMENTATION FROM TURBULENT FLOW

K. I. Heiskanen

Karelian Branch, USSR Academy of Sciences

ABSTRACT

Most interpretations on the origin of sedimentary rocks are based on the calculation of a few distribution moments of grain-size measurements. All information about the sedimentary environment contained in the granulometric spectrum is not completely used. Utilization of all information is possible only with a model describing grain-size distribution in the formation mechanism. The problem of construction of an adequate sedimentation model is separated into three principal parts: erosion, transportation and sediment deposition. Turbulent flow theory, flume experiments and river hydrologic data permit the creation of a corresponding mathematical model.

The erosion of bottom material by flow with normally distributed instantaneous velocities and pressures corresponds to the point-source diffusion scheme. Particle fall velocity of suspended material is in proportion to the probability of the moment velocities to exceed the turbulent intensity.

All three types of the particle transportation (rolling, saltation and moving in suspended state) are the result of the single chance-wanderment mechanism and are described by the limit form of diffusion in the gravity field.

The fall-velocities distribution obtained coincides with the A. Einstein's "sedimentary distribution". The particle leap height distribution is obtained in the same manner. Settling of suspended sediment is determined by the law of turbulent intensity change.

Sediment particle fall-velocity distribution is determined by the form of this law and parameters of "sedimentary distribution".

A general model operator transforming the particle fall-velocity distribution of the material washed into the distribution of the settled particles has a composite form even for the simple example. So the parameter estimation is accomplished by minimizing the mean square difference using any suitable method.

The suggested model is in good agreement with the vertical particle distribution observed in flume experiments and natural rivers. Settled sediment in the flows investigated has the same concordance. Thus, the approach may give a new tool for the reconstruction of some depositional environment characteristics.

INTRODUCTION

Many attempts have been made to use grain-size moment characteristics of sedimentary rocks to interpret their origin (Rukhin, 1947; Pettijohn, 1957; Passega, 1964; and many others). With few exceptions (Klovan, 1966), all information about the environment of sediment accumulation contained in granulometric data is not used. The critical-diagram fields are the main tools of the investigations and are based on flow, sediment-size composition measurements. However, it seems possible to approach the interpretation of sedimentary grain-size distribution analyses by modeling the processes responsible for the distributions. Also an attempt to construct the analytical description of some sedimentation processes to create an adequate model of this phenomenon should be undertaken. This paper deals with a simple scheme of sediment generation including erosion, suspension, transportation and deposition from turbulent flow.

EROSION AND SUSPENSION

It has been ascertained by numerous investigations that the instantaneous velocities and pressures of turbulent flow are normally distributed (Makkaveev, 1947; Karaushev, 1948). Thus it is natural to suppose that the relative quantity of particles drawn in motion by water currents is determined by the probability of increasing water-current velocities to exceed the particle fall-velocity values (Klaven, 1968). On the other hand, similar conclusions comes from the consideration of bed scour as the result of turbulent diffusion in the gravity field. If the distribution of particles diffusing from the bottom is in accordance with the diffusion equation (Feller, 1957),

$$\frac{\partial v(x,t)}{\partial t} = -2c\frac{\partial v(x,t)}{\partial x} + D\frac{\partial^2 v(x,t)}{\partial x^2}, \tag{1}$$

where

$v(x,t)$ = the probability of the particle to be at point x at the time moment t,

x = the distance from the bottom,

c = the current-fall velocity in this example and,

D = diffusion coefficient,

then the distribution-density function would have the form

$$f_t(x) = \frac{1}{\sqrt{4\pi Dt}} e^{-\frac{(x-2ct')^2}{4Dt}}, \quad x > 0.$$

The fall-velocity distribution of the suspended material corresponds therefore to the form

$$r(c,D) = \frac{2k}{\sqrt{4\pi Dt}} \int_0^\infty e^{-\frac{(x-2ct')^2}{4Dt}} dx.$$

The substitution y for $\frac{x-2ct}{\sqrt{2Dt}}$ reduces it to

$$r(c,D) = 2k\left[1-\phi\left(c\sqrt{\frac{2t'}{D}}\right)\right]. \tag{2}$$

This function $r(c,D)$ may be regarded as a weighting function, determining the transition of particles with different fall velocities in suspension. In particular (and it is in a good agreement with intuition), c is equal to 0 or ∞ and gives 1.0 or 0 according for this function. It means that the finest particles are suspended but the coarsest ones remain on the bottom. Even so, equation (2) is equal to 0 or 1.0 if the parameter D has the value corresponding to value 0 or ∞. These values correspond to the lack of erosion by laminar current and total washout of coarse material by the infinitely intensive turbulent streams. The weighting function $r(c,D)$ may be taken as a density function of the particle fall-velocity distribution, if the particle fall velocity has uniform distribution. If the bottom material submits to another density function, the suspended particle fall-velocity distribution would be in proportion with

$$w_\varepsilon(c,D) = g(c) \cdot r(c,D). \tag{3}$$

TRANSPORT OF SUSPENDED SEDIMENTS

Usually three types of sediment transport are considered: rolling, saltation (uneven moving of particles) and moving in a suspended state. In all situations the basic cause of particle movement is a random velocity and pressure pulsation. This is the nature of turbulence and the uneven chance-wandering of particles is similar to diffusion (Fidman, 1959; Mikhailova, 1966, Karaushev, 1948; Makkaveev, 1931). Experiments show this analogy to be real. Thus, this conclusion is the base of sediment-transport diffusion theory (Makkaveev, 1931; Velikanov, 1938).

Three types of sediment movement result in the same chance-wandering mechanism and differ only by various heights of particle raising. The difference between the three movement types is transitional (Mikhailova, 1966). Rolling may be considered as a limited form of uneven particle movement (saltation) with extreme low leaps, and suspension as a series of leaps during which a particle has no time to settle to the bottom.

Mikhailova (1966, p. 54, fig. 15) gives the result of measurements of particle leap heights obtained by flume experiments which collaborates the movement mechanism. The distribution-density function of the particle leap height may be estimated from the particle trajectory maximum density function which is known from diffusion theory (Feller, 1964),

$$f(z,t) = \frac{z}{2\sqrt{\pi D t^3}} \, e^{-\frac{(z+2ct)^2}{4Dt}} ,$$

where t = time,

z = maximum of the particle leap height,

c = particle fall velocity, and

D = turbulent diffusion coefficient.

The particle leap-height, density function is deduced then by the following normalization

$$f'(z,t) = f(z,t) \cdot \left\{ \int_0^\infty f\,(z,t)dz \right\}^{-1} ,$$

and searching a limit

$$\lim_{t \to \infty} f'(z,t) = \frac{c^2}{D^2} \, z e^{-\frac{c}{D} z} . \qquad (4)$$

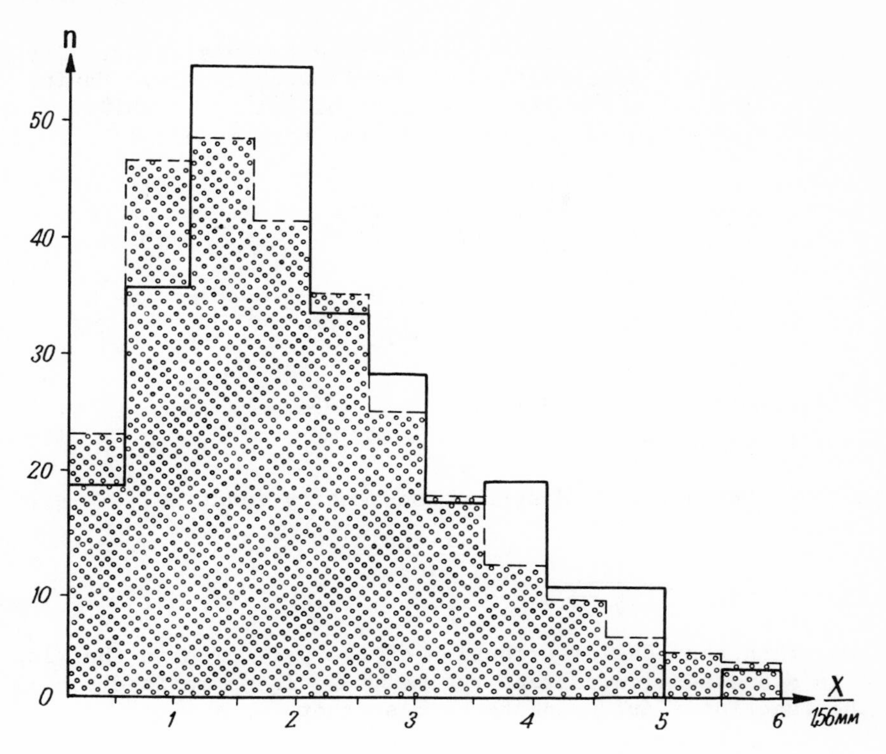

Figure 1. - Frequency curves of particle leap height distribution. Solid line, empirical data; dotted line, theoretical approximation.

So the particle leap-height, distribution-density function is a variety of the gamma-distribution density

$$h(z) = a^2 z e^{-az}, \quad z \geq 0.$$

For the mentioned experiments, this function is (Fig. 1),

$$h(z) = 0.0575 \cdot z \cdot exp(-0.2397 \cdot z).$$

A good agreement of this density function and experimental data is found by comparing $x^2 = 21.4; x^2_{0.01} = 23.2$. This result attests to the use of diffusion theory to a certain extent, although the agreement is not perfect (seemingly structural differences are involved in the nearbottom part of the flow in which turbulent the diffusion coefficient possibly depends on the distance from the bottom).

The equation of diffusion with the current gives opportunity to construct particle distribution in the flow thickness. Having the bottom as a material source gives the following probability of a particle to be between x_0 and x_1 at the time moment t:

$$P(x_0 \leq x \leq x_1) = \frac{1}{\sqrt{2\pi}} \int_{y_0}^{y_1} e^{-\frac{\lambda^2}{2}} d\lambda \,,$$

where
$$y_1 = \frac{x_1 - 2ct}{\sqrt{2Dt}} \,, \quad y_0 = \frac{x_0 - 2ct}{\sqrt{2Dt}} \,,$$

D = diffusion coefficient, and

c = fall velocity.

This leads to the particle distribution-density function,

$$f_t(x) = \frac{1}{\sqrt{4\pi Dt}} e^{-\frac{(x-2ct)^2}{4Dt}} \,.$$

It follows then that the distribution of the particles in the flow with depth H and settled current regime ($t \to \infty$) is normalized to avoid the function $f_t(x)$ going to zero so that

$$f_0(x;c,D) = \lim_{t \to \infty} \frac{f_t(x)}{\int_0^H f_t(x)dx} = \frac{c}{D} \frac{e^{\frac{c}{D}(H-x)}}{e^{\frac{c}{D}H} - 1} \,. \tag{5}$$

It is interesting to compare (5) with Einstein and Smolikhovsky's (1936) subcolloidal particle distribution about a horizontal screen in the field of gravity

$$f_E(x) = Ae^{-Bx} \,,$$

which can be termed "sedimentary distribution".

Diffusion theory of sediment material suspension suggested by Makkaveev (1931) gives the equation

$$s = s_c e^{-\frac{wh}{\varepsilon}\left(\zeta - \frac{c}{h}\right)} \,,$$

where s = turbidity,

s_c = turbidity for distance from the bottom,

w = fall velocity,

h = depth of flow,

ε = turbulent diffusion coefficient, and

$\zeta = \frac{y}{h}$ = relative depth of the flow.

So the distribution function deduced in (5) is

$$f_0(x;c,D) = \frac{c}{D} \frac{e^{\frac{c}{D}(H-x)}}{e^{\frac{c}{D}H} - 1} , \tag{6}$$

and corresponds to hydraulic theoretical and empirical data and to some mathematical models. The theoretical function for vertical distribution of suspended sediment (6) agrees satisfactorily with actual vertical distribution of suspended material in the Mississippi River at St. Louis (Jordan, 1965). Also an examination shows that the same values for the parameters D, H simultaneously give a good approximation for the vertical distribution of every fraction of material in the size range (Fig. 2). If all suspended material has a uniform particle fall-velocity distribution then (6) would be in proportion to the actual particle fall-velocity distribution at any distance above the stream bed. Equation (6)

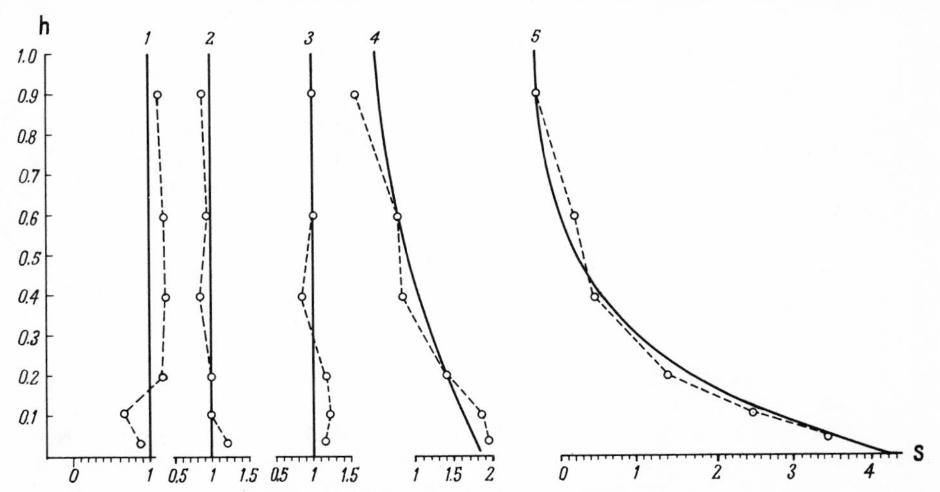

Figure 2. - Vertical distribution of relative turbidity of several (1-5) grade sizes. Solid line, theoretical; dotted line, empirical data.

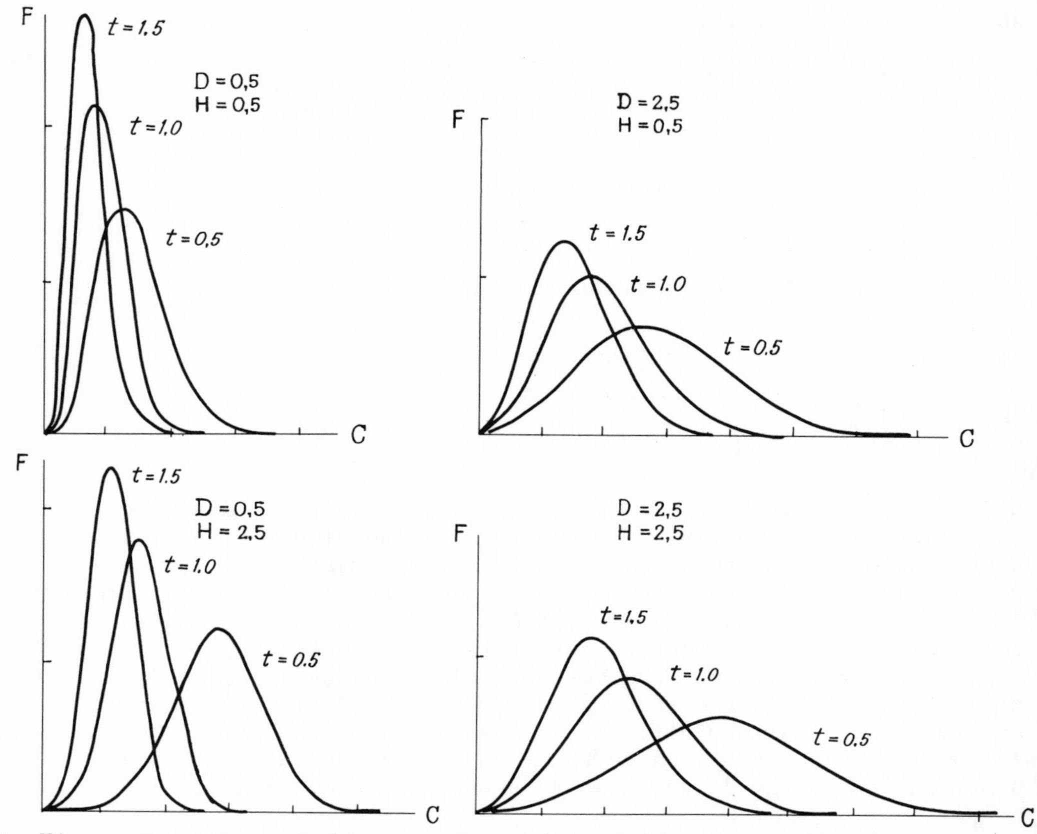

Figure 3. - General diagram of particle fall velocity distribution-density function.

however has to include a weighting function, which is normalized and must be the particle fall-velocity distribution density for the total suspended material. The same equation is for material carried away from the point of the flow erosion (3)

$$f(x,c;D) = k \cdot w(c,D) \cdot f_0(x;c,D). \tag{7}$$

SEDIMENT DEPOSITION

A two-dimensional fall-velocity, distribution-density function of particles suspended in the flow permits an approach to the search of a corresponding density function for material settled on the bed. It seems evident that slowing of the current, and hence the character of turbulency, may be accomplished by different methods. The simplest method practically is the momentary stop of flow. If at the moment $t = 0$, all suspended particles drop to the bottom with velocities equal to their fall velocities, then at the moment $t = t_1$, the particles with a distance of no more than $x = ct$, from the bottom would become bottom sediment. Therefore at any time moment $t > 0$, any part of the total turbidity equal to

$$\int_0^{ct} f(x,c;D)dx = \frac{1-e^{-\frac{c}{D}t}}{1-e^{-\frac{c}{D}H}} \cdot w(c,D)$$

is in the bottom sediment. The differential of this equation that characterizes the sediment settling at the current-time moment is the sediment particle fall-velocity, distribution-density function to the constant

$$f(c) = k \cdot w(c,D) \cdot \frac{e^{-\frac{t}{D}c^2}}{1-e^{-\frac{H}{D}c}} .$$

General shape of this function is shown on Figure 3.

Now the three basic parts of sedimentation, erosion, transportation and deposition, have been considered theoretically. The simplest models (2), (7) and (8) have been constructed for each of them. Before examination of the principal model (8), it is necessary to specify the form of the function $g(x)$ determining the particle grain-size distribution of the eroded material.

It is known that the distribution density of the particle size obtained by crushing corresponds to the exponential (Bennett, 1936) or lognormal (Kolmogoroff, 1941) law. The principal source of the natural clastic material is rock weathering. The form of the particle distribution obtained by rock weathering is not well known, therefore, it is approximated by the simplest form - the exponential law. This hypothesis is examined with quartz-grain distribution data obtained by the natural disintegration of the different rock types without any differentiation (Blatt, 1967). It is known that all rock types (massive plutonic rocks, average of 19 samples; gneiss, average of 16 samples; metamorphic schists, average of 6 samples) generate quartz grains with a size distribution which fits approximately the exponential law (Fig. 4). So the exponential law of the size distribution is accepted for the particles at the point of washout

$$g(x) = \lambda e^{-2x} . \tag{9}$$

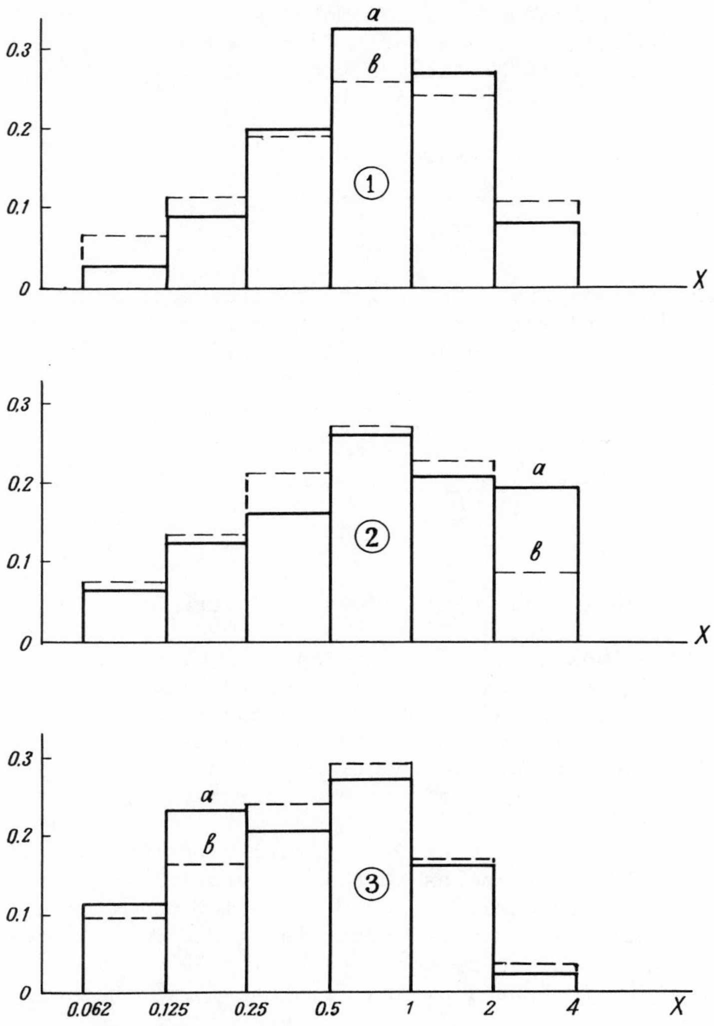

Figure 4. - Size-frequency curves of quartz grains obtained by weathering of massive plutonic rocks (1), gneiss (2) and metamorphic schists (3). Solid line, Blatt's (1967) measurements; dotted line, exponential approximation.

Substitution of the empirical polynom binding the particle size with its fall velocity transforms (9) into the function of the fall velocity

$$g(c) = \lambda \cdot exp\ [-\lambda P(c)]\ .$$

The empirical polynom $x = P(c)$ and the opposite $c = Q(x)$ are obtained by the least-square method with use of quartz-grain fall-velocity measurements (Sarkisian, 1958).

Finally the bottom sediment particle fall-velocity distribution-density function has the complete form,

$$f(c) = \frac{c^2\left[1-\phi\left(c\sqrt{\frac{2t'}{D}}\right)\right]}{\exp\left[\frac{t}{D}c^2-\frac{H}{D}c+\lambda P(c)\right]\left[e^{\frac{H}{D}c}-1\right]\int_0^\infty \frac{c^2\left[1-\phi\left(c\sqrt{\frac{2t'}{D}}\right)\right]}{\exp\left[\frac{t}{D}c^2-\frac{H}{D}c+2P(c)\right]\left[e^{\frac{H}{D}c}-1\right]}} \tag{10}$$

where

$$\phi(z) = \frac{1}{\sqrt{2\pi}}\int_{-\infty}^{z} e^{-\frac{u^2}{2}}\,du.$$

To test the function (10) with the natural sediment data, bottom-material measurements of the Polomet River (Razumikhina, 1967) were used. Test of the hypothesis (10) is accomplished with help of a special algorithm. The bottom-sediment size grades b_i $(i=0,\ldots,n)$ were transformed into the corresponding fall-velocity values by the polynom $c = Q(x)$,

$$c_i = Q(b_i).$$

Using the given grade size percentages $p_i(i=1,2,3,\ldots,n;\sum_1^n p_i=1)$, least squares gives the expression

$$W_n = \sum_{i=1}^{n}\left\{\int_{b_{i=1}}^{b_i} f(c)dc - p_i\right\}^2$$

Minimizing the function W_n by the fastest reducing method (anti-gradient), it is possible to determine the corresponding values of the variables D, H, t, λ. The calculations were made with a MIR-1 computer by V. L. Fainberg. Ten samples of the bottom sediments collected simultaneously with deep and shallow water velocity measurements were analyzed. Although the minimum values of the function W_n obtained are small $(0.02 \leq W_n \leq 0.005)$, the correlation between the measured values of the deep and shallow water velocity and the values of H and D was not significant. Obviously the sampled sediments were deposited by flow with other characteristics.

The second part of the technique deals with granulometric data obtained by Kuenen and Migliorini (1950). They experimented with turbidity currents. When a suspension of known composition was poured in a ditch of constant depth, it was deposited in layers. Analysis of the granulometry of each layer shows that their deposition corresponds practically to the constant values of t, H, λ parameters and decreasing ones of the diffusion coefficient D. For the example, samples collected at intervals 0-4mm, 8-10mm and 13-15mm from the bottom give correspondingly the next values of the turbulent diffusion coefficient: D = 41.14, 6.04, 2.21, whereas the other parameters are nearly constant: t = 48.2, 48.3, 49.9; H = 49.3, 49.2, 49.6; λ = 47.7, 47.0, 49.0.

Development of the model perhaps will give an approach to ancient flow characteristics by using granulometric data.

REFERENCES

Bennett, J. G., 1936, Broken coal: Jour. Inst. Fuel., v. 10, p. 22-39.

Blatt, H., 1967, Original characteristics of clastic quartz grains: Jour. Sed. Pet., v. 37, no. 2, p. 401-424.

Einstein, A., and Smolikhovsky, M., 1936, Brownian movement: United Scientific & Tech. Publ. House, People's Commissariat of Heavy Industry, Moscow.

Feller, V., 1964, Introduction to the study of probability and its application: World, Moscow.

Fidman, B. A., 1959, Some data on flow in the viscous layer of a turbulent stream, in Hydratic structures and the dynamics of river beds: Publ. Acad. of Sci. USSR, Moscow, p. 183-188.

Jordan, P. R., 1965, Fluvial sediment of the Mississippi River at St. Louis, Missouri: U.S. Geological Survey Water Supply Paper 1802, 89 p.

Karaushev, A. V., 1948, Computing the distribution of deposits in flows: State Hydrological Inst. Publ., sec. 8, no. 62, Hydrometeorological Publ. House, Leningrad, p. 40-80.

Klaven, A. B., 1968, The cinematic structure of a turbulent stream: State Hydrological Inst. Publ., sec. 147, Hydrometeorological Publ. House, Leningrad, p. 134-141.

Klovan, J. E., 1966, The use of factor analysis in determining depositional environments from grain-size distributions: Jour. Sed. Pet., v. 36, no. 1, p. 115-125.

Kolmogoroff, A. N., 1941, Uber das logarithmish normale Ferteilunggesetz der Dimensionen der Teilchen lei Zerstuckelung: Doklady Akad. Nauk SSSR, v. 31, 99 p.

Kuenen, Ph. H., and Migliorini, C. J., 1950, Turbidity currents as a cause of graded bedding: Jour. Geology, v. 58, no. 2, p. 91-127.

Makkaveev, V. M., 1931, On the theory of turbulent regime and weighing deposits: State Hydrological Inst. Bull., no. 32.

Makkaveev, V. M., 1947, The distribution of longitudinal and transverse velocities in open streams: State Hydrological Inst. Publ., sec. 2, no. 56, Hydrometeorological Publ. House, Leningrad.

Mikhailova, N. A., 1966, The transfer of solid particles by turbulent streams, Hydrometeorological Publ. House, Leningrad, 234 p.

Passega, R., 1964, Grain size representation by CM patterns as a geological tool: Jour. Sed. Pet., v. 34, no. 4, p. 830-847.

Pettijohn, F. J., 1957, Sedimentary rocks (2d ed.): Harper and Brothers, New York, 718 p.

Razumikhina, K. V., 1967, Native study and computing the transport of streams: State Hydrological Inst. Publ., sec. 141, Hydrometeorological Publ. House, Leningrad, p. 5-34.

Rukhin, L. B., 1947, The granulometric method of studying sands: Leningrad State Univ. Publ. House, Leningrad.

Sarkisian, A. A., 1958, The settling of deposits in a turbulent stream: Collection entitled "River Bed Processes", Publ. Acad. of Sci. USSR, Moscow, p. 338-351.

Velikanov, M. A., 1938, The bases of the statistical theory of weighing deposits: Meteorology and Hydrology, no 9-10.

CONDITIONAL SIMULATION OF SEDIMENTARY CYCLES IN THREE DIMENSIONS

J. Jacod and P. Joathon

Ecole Nationale Superieure des Mines

ABSTRACT

We propose a method to determine the most probable pattern of a sedimentary cycle between boreholes. The purpose of this conditional simulation method is to solve a three-dimensional quantitative problem. It differs essentially from classical methods of quantitative studies by involving several dimensions and by not using any Markovian property. The example is the Permo-Triassic sedimentary cycle of the Chemery structure in the Paris Basin. It is a fluvial or deltaic sequence of clayey and sandy lenses. The structure is outlined by 15 boreholes.

Some genetic hypotheses are made about the cycle. A random-genetic model is proposed which fits the available data. The most probable lithological correlations between layers of different boreholes are determined. To do this a maximum likelihood method is used. The most probable shape of lenses between boreholes for each group of contemporary lenses is determined. In this manner the problem is thus solved. The results are given as a set of thickness maps for successive groups of lenses and as cross sections.

INTRODUCTION

Quantitative properties of sedimentary structure, for example, between boreholes is not easily interpreted unless the cyclic sedimentation phenomenon is uniform. Several deterministic quantitative models of cyclic sedimentation exist, however, but they are used to describe the process of sedimentation, to determine genesis, and possibly to extrapolate trends. Because of the complexity of data, they must be summarized, and consequently do not allow prognostication.

The use of random models also has been explored. These models better represent sedimentary cycles where a large number of data are available or local irregularity is characteristic. Unfortunately, random models may have two major inadequacies (Krumbein, 1967; Schwarzacher, 1969). Firstly, they are purely descriptive, excluding any genetic consideration. Secondly, they are unidimensional. These two facts prohibit making use of them for prediction as emphasized by Krumbein and Dacey (1969). Yet some models are free from these constraints, such as the transgression and regression model of Krumbein (1968) and the random-genetic models of Jacod and Joathon (1971a).

The aim of this paper is to show that random-genetic models permit a more satisfactory description of cyclic sedimentation. More precisely, suppose that a model fits a sedimentary structure, then, we show that it is possible to simulate this structure conditionally to data collected along each borehole. We give an interpretation of the structure, which fits the experimental data, and which is the most probable form of the sedimentary process.

This paper is theoretical, and we do not wish to describe a particular sedimentary structure. However, we employ a specific model, the structure of Chemery (also called Contres) in the Paris Basin. This structure was selected for the study because it has been used for underground gas storage since 1967 by the "Gaz de France" Company.

In the first part of the paper, we describe the structure, the genetic hypotheses and the model itself (Jacod and Joathon, 1971b). Then, we set forth the meaning of conditional simulation, and give a detailed account of the proposed method. And lastly, we furnish some results of simulation using the Chemery structure as an example.

With the tools that are employed here, we will be able to solve other problems by conditional simulation. This is especially true for quantitative sedimentary problems.

THE RANDOM-GENETIC MODEL

Sedimentary Structure of Chemery

The Chemery structure is located near the village of Chemery in the Paris Basin (Loir-et-Cher). We studied a part of the Permo-Triassic interval, which consists of clay and sandstone in approximately equal proportions. Overlain by Levallois' clays, it is favorable for underground gas storage. When the study began, the structure was outlined by 15 boreholes, 4 of which were not complete. Mean thickness of the interval is 45 m and horizontal proved area is about 25 km^2 (Fig. 1).

The structure consists of an extremely complex succession of clayey and sandy lenses (a mean of 36 lenses along a vertical line). Each lens is fairly well delimited. Tests of gas-flow and difficulties in lithological correlations indicated that the dimensions of the lenses were small (smaller than the minimal distance between boreholes). Geologically, these units originated as fluvial or deltaic sediments. The sediment source was probably from the south. Finally, an analysis of logs allows lithological correlations to be made although some correlations are somewhat uncertain.

Random-Genetic Models

To solve the problem which was discussed in the introduction (to develop a conditional simulation), we must have a model which

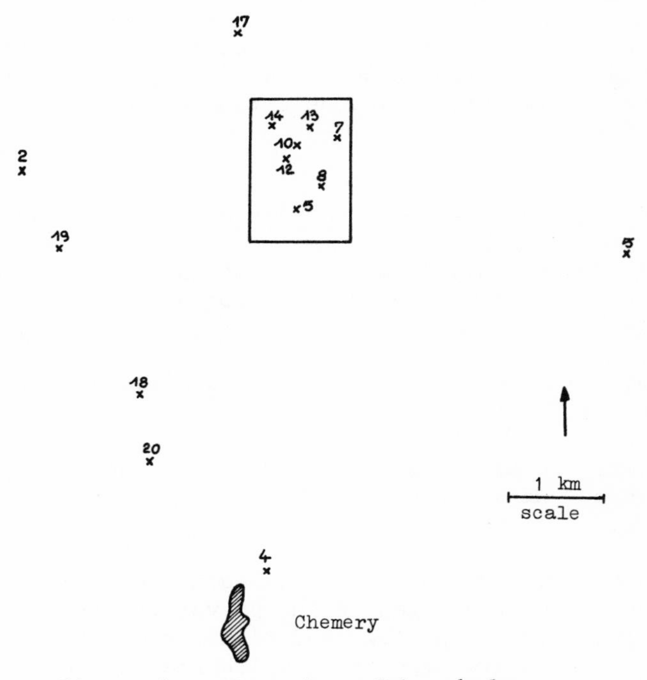

Figure 1. - Location of boreholes.

fits the sedimentary structure as well as possible. Indeed, only a model may allow us to interpret between boreholes;

This model must satisfy several conditions:

(1) It must be tridimensional;

(2) It must be sufficiently general to allow the introduction of new data; and

(3) It must be sufficiently detailed to realistically represent the structure, so that the simulation is meaningful.

On one hand, the number of data may be large (for Chemery, we know the thickness of 503 lenses), and such a large quantity of data has to be treated statistically. On the other hand, many factors affect the process of sedimentation, and although many have little influence, as a whole they cannot be neglected. For these two reasons, we use a random model.

A purely statistical model, however elaborate, omits important quantitative geological or genetic information. In fact it is this information which specifies the sedimentary structure and gives it some important features. Consequently we think it is desirable to use a genetic hypotheses as a framework for the model, and to adjust the model to the real situation. These genetic hypotheses give an explanatory value to the model, and not just a descriptive one. Moreover, they make a realistic tridimensional development easier.

Genetic Hypotheses For Chemery Structure

We are interested only in spatial distribution of lithologies. Thus we make the following simplifying hypothesis: lithologies are arbitrarily divided into two types, sandstone and clay.

Additional information concerning the genesis of the Chemery structure is necessary for the model.

(1) In order to maintain a shallow depth, there must be subsidence. We suppose that the subsidence rate is a constant.

(2) The chaotic appearance of the sediments is due to sudden and disordered changes in the prevailing flow conditions. According to given flow conditions, some clayey and sandy lenses are deposited, which are supposedly contemporary. As sandstone usually deposits faster than clay,

we may assume that in situation of cover, clay overlies sandstone. All the sandy lenses deposited according to given flow conditions form a sandy layer, and likewise for the clay. These layers are "lacunary", and may partly cover one another, then the clay part overlies the sandy part. When prevailing flow conditions suddenly change, two other layers are deposited, and so on.

(3) Let us suppose that new prevailing flow conditions involve noticeable erosion. One or several whole layers may be eroded. (This is actually a rough approximation.) On one hand, the model is simpler than reality but, on the other hand, taking the actual erosion into account would not cause important changes in the appearance of the simulation model.

We now have to specify what are the sedimentation factors which we retain. There are four principal factors and we neglect all others. First, subsidence which is characterized by a constant rate; as time is considered a proportional constant, we can give this rate the value 1. Next, sedimentation rate, which depends on time t and on the vertical line (represented by a point x of horizontal plane); this rate is characterized by a growing function $X_t(x)$, and sediment thickness during Δt interval is proportional to

$$\Delta X_t(x) = X_{t+\Delta t}(x) - X_t(x).$$

Then, the lithology that is deposited (clay or sandstone). And last, the depth $Y_t(x)$ at time t, along the vertical passing through x. Although not important from a geological point of view, depth plays a critical part when we study thickness, for it cannot be negative and depth must be considered in every quantitative model which describes a sedimentary process.

A second hypothesis leads us to admit that sedimentation is instantaneous; it takes place at successive times $T_1, T_2, \ldots T_n$. Moreover, at times with odd numbers, the deposit is sandy; at times with even numbers, it is clayey. For each x, the function $X_t(x)$ in t is constant in intervals (T_{n-1}, T_n). It jumps at times T_n (deposit of a lens along the vertical line passing through x). As layers are lacunary, some jumps may be zero. The appearance of the function $X_t(x)$ is shown on Figure 2. The associated succession of lithologies along the vertical line is (from bottom to top):

$$\text{sand - clay - clay - sand - clay.} \qquad (1)$$

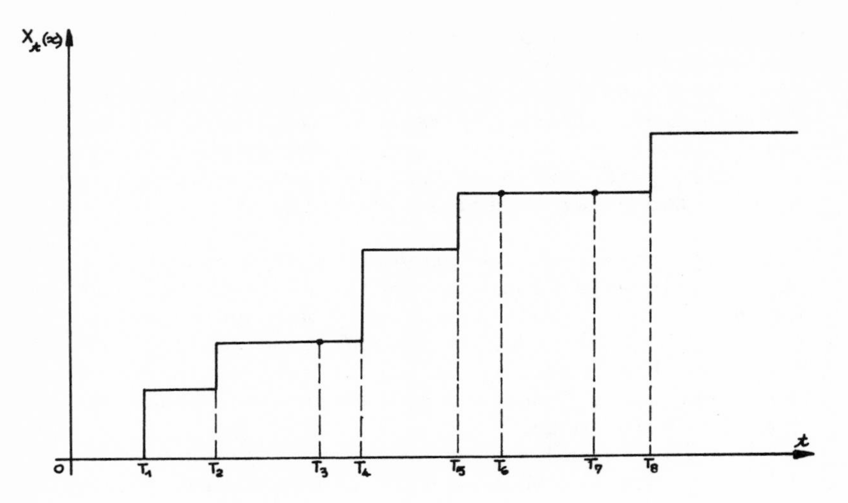

Figure 2. - Function $X_t(x)$.

The times T_n and the amplitude of jumps $\Delta X_n(x)$ at times T_n are random variables, the law will be specified later. This scheme seemingly ignores erosion. In fact it is compatible with erosion acting according to a third hypothesis. This type of erosion is considered by modifying the probability laws of variables T_n.

Now we suppose that subsidence, sedimentation rate and depth are connected by the following equation

$$dY_t(x) = dt - Y_t(x) \quad dX_t(x) \tag{2}$$

(t-differentials). In particular, this equation expresses that deposited thickness is proportional to depth. We see in this relationship a formalization of the following fact: the shallower the depth, the smaller the thickness.

Probability Laws

Now we will give the laws of random variables T_n and random functions $\Delta X_n(x)$. These laws are not arbitrary, in fact they were found through an analysis of experimental data.

The lengths $T_n - T_{n-1}$ will be independent random variables with the same exponential law, and with parameter γ

$$P(T_n - T_{n-1} < t) = e^{-\gamma t} \tag{3}$$

(see for example Feller, 1966). For Chemery we have $\gamma = 1.54$.

Concerning the sedimentation rate, specification is far more difficult. The different layers are probabilistically independent. Each layer is constituted of lenses constructed as follows: we consider first a Poissonian distribution of points on an horizontal plane. This distribution has a density θ_S for a sandy layer, θ_C for a clayey one. Around each Poissonian point I taken as a center, we put a parabolic cap with an elliptic base. The dimensions and the orientation of the basic ellipses are constant. The height of the parabolic cap at its center is a random variable Z, the distribution function of which is denoted by $N(z)$,

$$P(Z < z) = N(z). \tag{4}$$

In the example of Chemery, we take $\theta_S = 3.4\ 10^{-5} m^{-2}$, $\theta_A = 6.10^{-5}\ m^{-2}$. For basic ellipses, the ratio major axis/minor axis is 3/2. The major axis is oriented north-south, its length is 200 m for a sandy layer, 150 m for a clayey one. The exact mathematical formulation of $N(z)$ is too detailed to be given here, and we will be satisfied with giving its representative curve (Fig. 3).

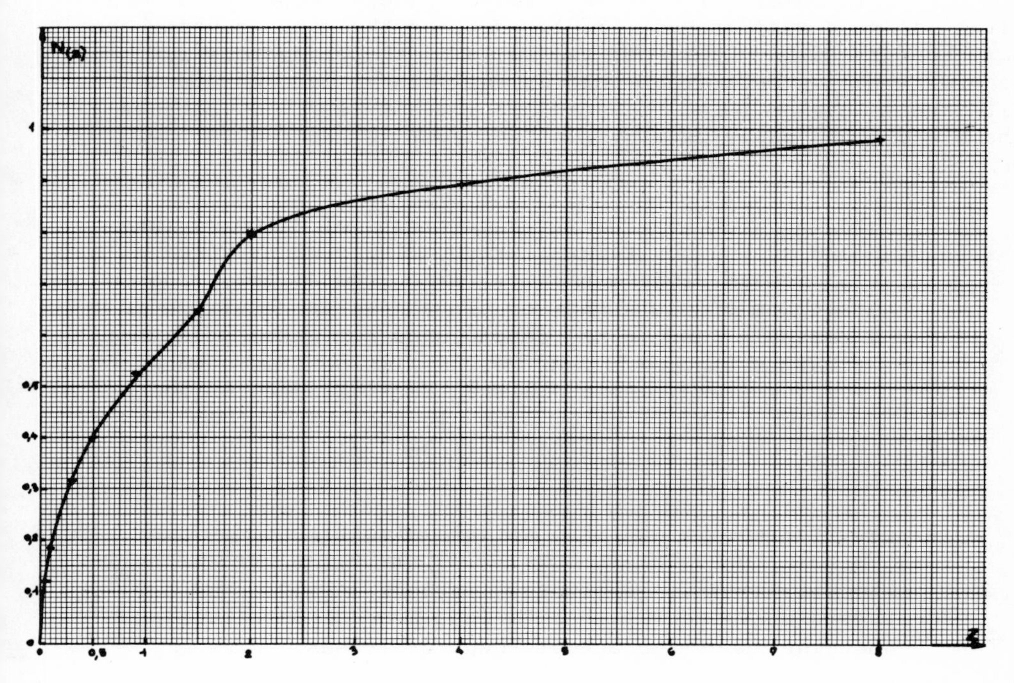

Figure 3. - Representative curve of function $N(z)$.

We now are able to define $\Delta X_n(x)$. If point x is not located in any basic ellipse, we set $\Delta X_n(x) = 0$. If x is in one or several basic ellipses, we take for $\Delta X_n(x)$ the largest height of the corresponding parabolic caps through point x.

This completes the specification of the probability laws of the model. In particular, we may compute the probability β of having $\Delta X_n(x) = 0$; a priori this probability depends on layer lithology, but the two values for sand and clay are assumed to be equal (for Chemery, we find $\beta = 0.52$). We also can determine the law of the jump $\Delta X_n(x)$, when this jump is not zero. The function $N(z)$ has been chosen in order for this law to be a so-called gamma law, with parameters b and δ,

$$P[X_n(x) < y] = F(y) = \frac{b}{\Gamma(\delta)} \int_0^y e^{-bz} z^{\delta-1} dz, \tag{5}$$

for Chemery, $b = \delta = 0.5$.

Testing the Model

The principles for testing a random-genetic model have been given in Jacod and Joathon (1970). Neither the genetic hypotheses, nor the form of the laws, nor the value of the parameters are haphazard. We have determined them by analysis of available data. All details can be found in Jacod and Joathon (1971b). Because the aim of this paper is different, however we add the following remarks.

(1) The qualitative properties of data (for example, the shape of the covariogram or covariance of successive thicknesses along a vertical line, the shape of the histogram of the number of clayey lenses occurring in a borehole, and so on) depend upon the genesis of sedimentary structure. Thus we can eliminate some genetic hypotheses and retain others. For instance, a random arrangement of lithologies could not be accepted, and we had to assume an alternate arrangement of clay and sandstone.

(2) For the choice of laws, we have almost no constraint. We take the handiest ones. In particular for the horizontal extension of lenses, we have assumed somewhat oversimplified characteristics, because we know almost nothing about the actual shape of lenses.

(3) The values of the parameters $(\beta, \gamma, \delta, b)$ are obtained from a statistical study of the data. In the example of Chemery (shallow depth, and comparatively large number of layers), they are known fairly precisely.

(4) Finally, we have more numerical relations than parameters to be estimated. These relations are compatible with each other, which constitutes a criterion of fitness for the retained model.

PRINCIPLES FOR CONDITIONAL SIMULATION

When we have a random model of sedimentation, a simulation according to this model can have several meanings. It can be a plain simulation; we choose each of the random variables according to its own law. This allows the step-to-step construction of the sedimentary structure. It also can be the most probable simulation; we obtain reality which maximizes the probability with the help of a maximum likelihood method. Finally, we can combine these two possibilities, for certain conditions.

Now, let $(Y_1 \ldots Y_n)$ be a family of random variables; the knowledge that some variables influence the probability law of others, makes it necessary to consider the conditional probability law. We mean conditional simulation in this sense. We know the thickness of lenses occurring in each borehole (a priori, they are random variables). The probability law of unknown random variables or functions (i.e. shape and dimensions of lenses between boreholes) is then a conditional probability law. A conditional simulation is a simulation (plain, or the most probable), according to this conditional law.

A conditional simulation has two chief advantages:

(1) A simulation depends on the model, which itself depends only on statistics computed and thus there is a great loss of information. A conditional simulation takes individual data into consideration and all information is used.

(2) A simulation has almost no chance to fit exactly the experimental data but a conditional one does. However, we must emphasize that we obtain the probable shape of structure, and not the exact one.

Our first task then is to find the conditional law. This problem seems easy, because if we know the laws of the model, we can (theoretically) find the probability of any event. Unfortunately, it is not so simple. Indeed, the sedimentary structure depends on a random number of random variables (number of layers; for each layer, location and thickness of each cap; intervals between the deposits of two layers). These variables are a priori independent, but if we know thickness in each borehole, they are dependent. We clearly cannot handle this number of variables unless some approximations are made.

$(Y_1 \ldots Y_n)$ was the previous set of random variables, so let us denote by F_p the law of $(Y_{p+1}, \ldots Y_n)$ conditionally to $(Y_1, \ldots, Y_p)$. and by G_p the law of Y_{p+1} conditionally to $(Y_1, \ldots, Y_p)$. It is well known that F_p can be expressed easily in the form of G_p, G_{p+1}, $\ldots, G_{n-1}$. This fact allows the step-by-step progression in the determination of the conditional law.

Before going further, let us explain some terminology. Along a vertical line, the deposit of a layer is called a lens. Data are divided into two parts: (1) the resultant configuration, i.e. the succession of lithologies in each borehole and, possibly, lithological correlations; and (2) the thickness of each lens. Likewise, we must determine the conditional law of two different types of random variables. Firstly, the actual configuration in each borehole, i.e., in fact for each layer to be blank or not blank along each one. Secondly, for each layer, the number of parabolic caps (which are called microlenses), their location and their height. Notice that knowing the actual configuration in each borehole implies the knowledge of all lithological correlations between boreholes. In a borehole, we call macrolenses a succession of lenses with the same lithology, ended at each side by a lens of another lithology.

To solve our problem, we take two major steps.

Step 1 - Actual configuration in each borehole. It depends on the resultant configuration. It also depends on the thickness of lenses.

(i) If the diameter of a microlens is large, the corresponding jumps of $X_t(x)$ have almost similar values for adjacent boreholes. Two thick lenses occurring in adjacent boreholes have a good chance to belong to the same microlens, and so to the same layer. For Chemery, we eliminate this possibility by assuming the diameter of microlenses to be less to the minimal distance between boreholes.

(ii) The thickness of a lens depends on depth. Now, the greater is the number of successive blank layers along a vertical line, the larger is the depth, because of subsidence. Therefore, a thick lens has more chance to follow several blank layers than a thin one.

We will neglect the dependence of the actual configuration and the thickness. For us, the actual configuration will depend only upon the resultant one. So we make an approximation, the range of which is difficult to estimate. But this approximation is necessary in order to solve the problem.

Step 2 - Microlenses of each layer. When the first step is achieved, we know the thickness of each layer in each borehole. The second step is easy enough to solve.

Now, how to simulate? If we want a plain simulation, we can follow a step-by-step procedure, which follows exactly the procedure of determination of the conditional law. When we simulate many random variables, we have a good chance of getting several aberrant values. But in a step-by-step procedure, if one of the first variables takes an aberrant value, we then introduce a huge bias, and the result has a poor significance.

On the other hand, a step-by-step procedure for the conditional law is not at all adapted to a maximum likelihood method. For this one we must dispose of the complete law. Then we cannot determine the exact most probable sedimentary structure. Thus we propose the following method. The first steps consist of finding a set of variables which have different conditional laws; we use the maximum likelihood method for these steps. The last steps consist of determining several sets of many variables, having the same law; for them, we make a plain simulation. This method avoids both obtaining aberrant values at the beginning of the procedure, and an illusory regularity which we would obtain if using the maximum likelihood method, successively for many variables having the same law; we then get the same value for each one. Yet we do not obtain exactly the most probable structure, but only a (somewhat bad) approximation of it.

RESEARCH FOR CONDITIONAL LAWS

Actual Configuration in Boreholes

The actual configuration depends only on the resultant one. To start with, we suppose that there is no lithological correlation.

Notations. By convention, we assume that in each of the boreholes, the first macrolens is sandy (with zero thickness if the first lens is clayey), and the last macrolens is clayey (with zero thickness if the last lens is sandy). Thus the number of macrolenses is always even.

The resultant configuration in each borehole is characterized by the number A of couples of macrolenses and the numbers m_i and n_i of lenses into the i-th sandy and clayey macrolenses (m_1 and n_A may be zero).

The actual configuration is characterized by the number A and the number Q of couples of layers, which does not depend on the boreholes; the number B_j of sandy (resp. C_j of clayey) layers between and including the first lens of the j-th sandy (resp. clayey) macrolens and the first lens of the j-th clayey (resp. $(j+1)$-th sandy) macrolens. In this description for each macrolens we have placed the m_j or n_j nonblank layers among the B_j or C_j possibilities.

We set $D_j = B_j + C_j - 1$. For each borehole we obtain

$$Q = \sum_{j=1}^{A} D_j \tag{6}$$

$$\begin{aligned} D_j &\geq m_j + n_j & & \\ D_1 &\geq 1 + n_j & \text{if } m_1 &= 0 \\ D_A &\geq m_A + 1 & \text{if } n_A &= 0 \end{aligned} \tag{7}$$

Figure 4 illustrates these definitions.

Estimation of Q. Notice that some difficulties arise in the determination of Q. The model describes the sedimentation process, but not its end, and Q measures the duration of the process. That is why we do not give the conditional law of Q. It would be meaningless. We just give an estimation of Q. Q is truly a parameter of the model.

On one hand, equations (6) and (7) provide a minimal value Q_o^i of Q for each borehole i. On the other hand, if N is the mean number of lenses along a vertical line, $N/2\beta$ is an estimator of Q. Thus we take for Q the maximum (integer) value of $N/2\beta$ and each Q_o^i.

Conditional Law of Actual Configuration. Here, we only discuss the method; all mathematical formulae which are complex, are included in the Appendix.

(1) We note that the actual configuration along each borehole depends only on the resultant configuration in that borehole. It does not depend on the actual or resultant configuration along other ones.

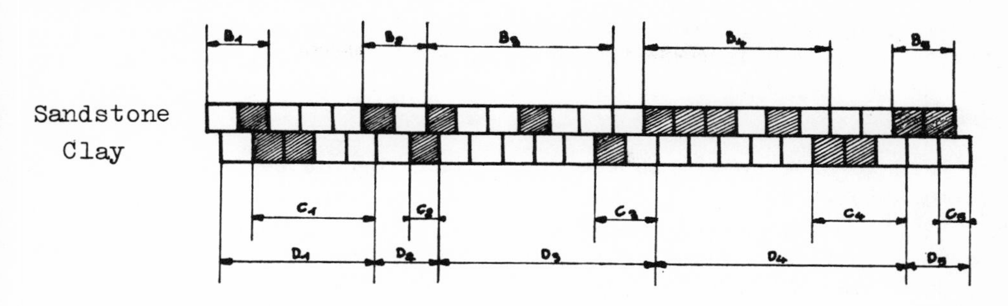

Resultant configuration :

$A = 5$

$m_1 = 1$	$m_2 = 1$	$m_3 = 2$	$m_4 = 4$	$m_5 = 2$
$n_1 = 2$	$n_2 = 1$	$n_3 = 1$	$n_4 = 2$	$n_5 = 0$

Actual configuration :

$Q = 24$

$B_1 = 2$	$B_2 = 2$	$B_3 = 6$	$B_4 = 6$	$B_5 = 2$
$C_1 = 4$	$C_2 = 1$	$C_3 = 2$	$C_4 = 3$	$C_5 = 1$

Figure 4. - Example of actual configuration, with corresponding resultant one. Each square represents layer. Blank square is blank layer.

(2) For a fixed borehole, we first determine the law of D_1, which depends only on Q and $(m_i, n_i)_{1 \leq i \leq A}$. Then the law of D_2, which depends on $Q - D_1$ and $(m_i, n_i)_{2 \leq i \leq A}$, and so forth until D_{A-1}. At last we have

$$D_A = Q - \sum_{j=1}^{A-1} D_j .$$

(3) We find that B_i and C_i depend on D_i, m_i and n_i, but not on D_j, m_j or n_j, with $j \neq i$. We determine the B_1 law, and we obtain C_i by $C_i = D_i + 1 - B_i$.

(4) We must place the m_i nonblank layers of the i-th sandy macrolens along B_i possibilities. This operation depends only on B_i and m_i. Unless $i = 1$, there must be a lens at the first place (the lowest one). The other ones must be independently set among

the remaining places, each place having the same probability, different lenses being in different places. We do the same for clay layers, except that for $i = 1$, the first place also must be occupied.

Thus we have developed the conditional law of the actual configuration. As to the simulation, we use maximum likelihood for step 2, plain simulation for steps 3 and 4.

Lithological Correlations. Let us now assume that there are lithological correlations. A correlation between different lenses means that these lenses belong to the same layer. Thus the actual configuration must include the number of the layer corresponding to each correlation.

The minimal value of Q is no longer the maximum of Q_o^i. The existence of correlations may increase this value. Its determination is a simple logical problem. For Chemery we find a minimal value equal to 41 if we take the correlations into consideration, to 38 if we do not.

The four previous points must be preceded by a preliminary step; the determination of the number of each correlation layer. This problem cannot be divided into several steps, and the method is given in the Appendix. For the simulation, we use the maximum likelihood method. When the number of each correlation layer is known, the correlations divide each borehole into several zones, the length of each one being known (it is in fact a certain number of layers). We deal with each zone as with the boreholes in the previous section, where we substitute the zone length to the number Q.

Microlenses

Now we assume a known actual configuration. Thus for each layer, we know the thickness (possibly zero) along each line of boreholes. We call U_n^i the thickness of the n-th layer upon drilling i, and ΔX_n^i the corresponding jump of the function $X_t(x)$.

Law of ΔX_n^i. The thickness U_n^i depends on ΔX_n^i, and also on the depth. Although we have no direct quantitative information about the depth, we must take it into account to find ΔX_n^i. We denote by Y_n^{i-} and Y_n^{i+} the depth along drilling i, just before and just after the deposit of the n-th layer, and by S_n the time interval between the $(n-1)$-th and the n-th layer: $S_n = T_n - T_{n-1}$. We get

$$Y_n^{i-} = Y_{n-1}^{i+} + S_n \tag{8}$$

$$Y_n^{i+} = Y_n^{i-} - U_n^i \tag{9}$$

$$\Delta X_n^i = -\,\mathrm{Log}\left(1 - \frac{U_n^i}{Y_n^{i-}}\right) \tag{10}$$

Consequently the knowledge of the Y_n^{i-} laws suffices to know the law of ΔX_n^i, according to equation (10). We purpose the following method.

(1) We start with a constant depth Y_1^{i-}, equal to the mean depth before a deposit. This constant can be computed with the model parameters. For Chemery, it is 2.5 m From equation (9) we deduce Y_1^{i+}.

(2) We look for the law of S_2, conditionally to Y_1^{i+} and U_2^i along each drilling. From equation (8) and (9), we deduce Y_2^{i-} and Y_2^{i+}. Then we notice that the law of S_3, conditionally to Y_2^{i+} and U_3^i is the same as the previous law. We deduce Y_3^{i-} and Y_3^{i+} and so on. Thus we have a method giving us the law of Y_n^{i-}, and of ΔX_n^i by equation (10).

In the first step, we introduce an approximation, the initial depth is not really a constant one, in fact it is a random function. When depth is of the same order of magnitude as the thickness, as for Chemery, we know (Jacod and Joathon, 1970) that this hypothesis does not introduce important errors, except for the first layers. When depth is large, we can consider it as a constant, with a fairly good approximation. In this situation we can avoid any reference to steps 1 and 2, because equation (10) gives us directly the values of ΔX_n^i.

In the second step, we assume that the conditional law of S_n, when Y_p^{i+} for $p \leq n-1$ and U_p^i for any p are known, depends only on Y_{n-1}^{i+} and U_n^i. In fact it does not depend on Y_p^{i+} for $p \leq n-2$, nor on U_p^i for $p \leq n-1$, but it depends on U_p^i for $p \geq n+2$ Here we introduce an approximation, but we are not able to estimate its range.

The conditional law of S_n is given in the Appendix. For these steps, we make a plain simulation.

Law of $\Delta X_n(x)$. Now we know the ΔX_n^i. It is clear that the random function $\Delta X_n(x)$ depends on ΔX_n^i, and not on ΔX_m^i with $m \neq n$. Thus we can treat each layer independently from the others, with the same method.

Around each borehole we define an influence zone, which is equal to the basic ellipse of the microlenses. Outside these zones, the knowledge of the ΔX_n^i does not yield any information about the process. The law of the microlenses outside the influence zone of the boreholes is the law specified in the section on probability laws.

Because of the hypothesis concerning the diameter of microlenses, we know that two influence zones cannot intersect. What happens in the zone of drilling i depends only on ΔX_n^i. The conditional law is given in the Appendix.

So we have developed the determination of the conditional law of the sedimentary structure. Likewise, the conditional simulation will be achieved when we have specified that we make a plain simulation for this last step. Indeed, starting with a constant initial depth, we have all the elements necessary to build the sedimentary structure.

SOME RESULTS

We have explained the theoretical scheme of the conditional simulation. It seems confusing, and moreover, the mathematical formulae are fairly complex. That is why it is important to know if the scheme is operational, i.e. if we really can obtain a conditional simulation.

A first remark. Each one of the many steps consists of a small number of simple operations. In particular, the most complex formulae define some probabilities by recurrence methods. Consequently, we can easily program this scheme. We use an IBM 360/40 computer with a 256 K memory for the computations and the simulations.

For the Chemery structure, 4 of 15 boreholes are incomplete, which introduces some extra complications -503 lenses, 12 lithological correlations, each of which connects a mean of 4 lenses. As an example, we give the successive thickness and lithology in two boreholes (No. 7 and 8, Table 1).

The first step, simulation of the numbers of the layers corresponding to correlations, takes about 10 min of computing. The second step, simulation of the actual configuration along the 15 boreholes, takes about the same time. The results are too many to be given extensively, so we only indicate the results for two boreholes (Table 2). This table may be compared with the first one.

To interpolate between boreholes, we draw a grid on a horizontal plane and simulate the thickness of each layer for each mesh point of this grid. The computing time is proportional to the number of mesh points. For 60 points and 82 layers, it takes less

Table 1. - Successive thickness and lithologies for units in boreholes No. 7 and 8.

Borehole No. 7

No. of lens	Lithology	Thickness	No. of lens	Lithology	Thickness
1	C	.6	23	C	.4
2	S	.6	24	C	.2
3	C	1.6	25	S	1.8
4	C	.6	26	S	1
5	C	2.2	27	S	2.2
6	C	.8	28	S	.8
7	C	.6	29	S	2.2
8	S	.2	30	S	3.6
9	C	.8	31	C	.8
10	C	1.4	32	C	1.8
11	C	1.6	33	S	3.4
12	C	.6	34	S	.4
13	S	.6	35	C	.2
14	C	.4	36	S	.4
15	S	.8	37	S	4.2
16	S	.6	38	C	2.6
17	C	.8	39	S	.4
18	S	.2	40	S	.4
19	C	.2	41	C	.4
20	S	2	42	S	.2
21	C	1.2	43	C	1
22	S	.6			

Borehole No. 8

No. of lens	Lithology	Thickness	No. of lens	Lithology	Thickness
1	C	1.2	20	S	1
2	S	8.4	21	C	1.6
3	S	.6	22	C	2.4
4	C	.8	23	C	.6
5	S	.6	24	C	1.6
6	C	.4	25	S	2.2
7	C	.4	26	C	1
8	C	1.6	27	S	1.2
9	S	.4	28	C	1
10	S	.2	29	S	1.8
11	S	.6	30	C	1
12	C	1.4	31	S	.8
13	S	1	32	C	.8
14	S	2.4	33	S	1.2
15	C	1.4	34	C	1
16	S	.6	35	C	.4
17	C	.8	36	C	.6
18	C	.6	37	S	2.6
19	C	.8			

Table 2. - Actual configuration along drillings No.7 and 8.

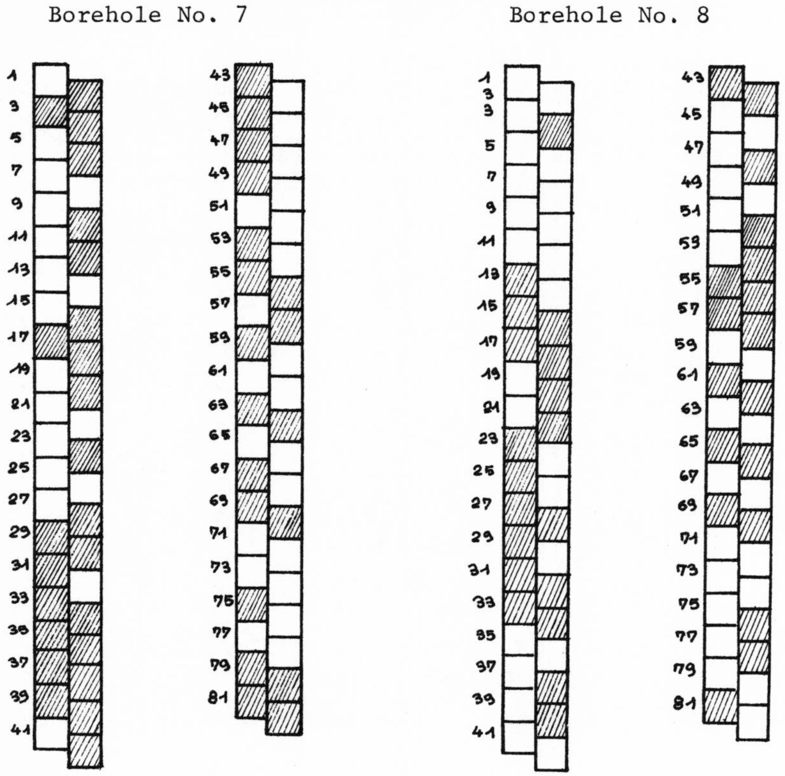

Borehole No. 7 Borehole No. 8

than 1 min.

We have simulated for a grid of 3844 points. It includes only a part of the structure about 1.5 km^2 in the best controlled area, so that we have 7 boreholes in the simulation area. We present the results as a set of maps. Each map is the thickness map of a layer. Thus we get an atlas of 82 maps, which fully describe the sedimentary structure in the simulated area. As all these maps have the same appearance, we only present one of them here (Fig. 5). To obtain a better visualization of the simulated structure, we also have drawn a north-south section, passing through the two boreholes (Fig. 6).

We can make several remarks about these results. The reader may verify that the simulation (Fig. 6) fits the exact experimen-

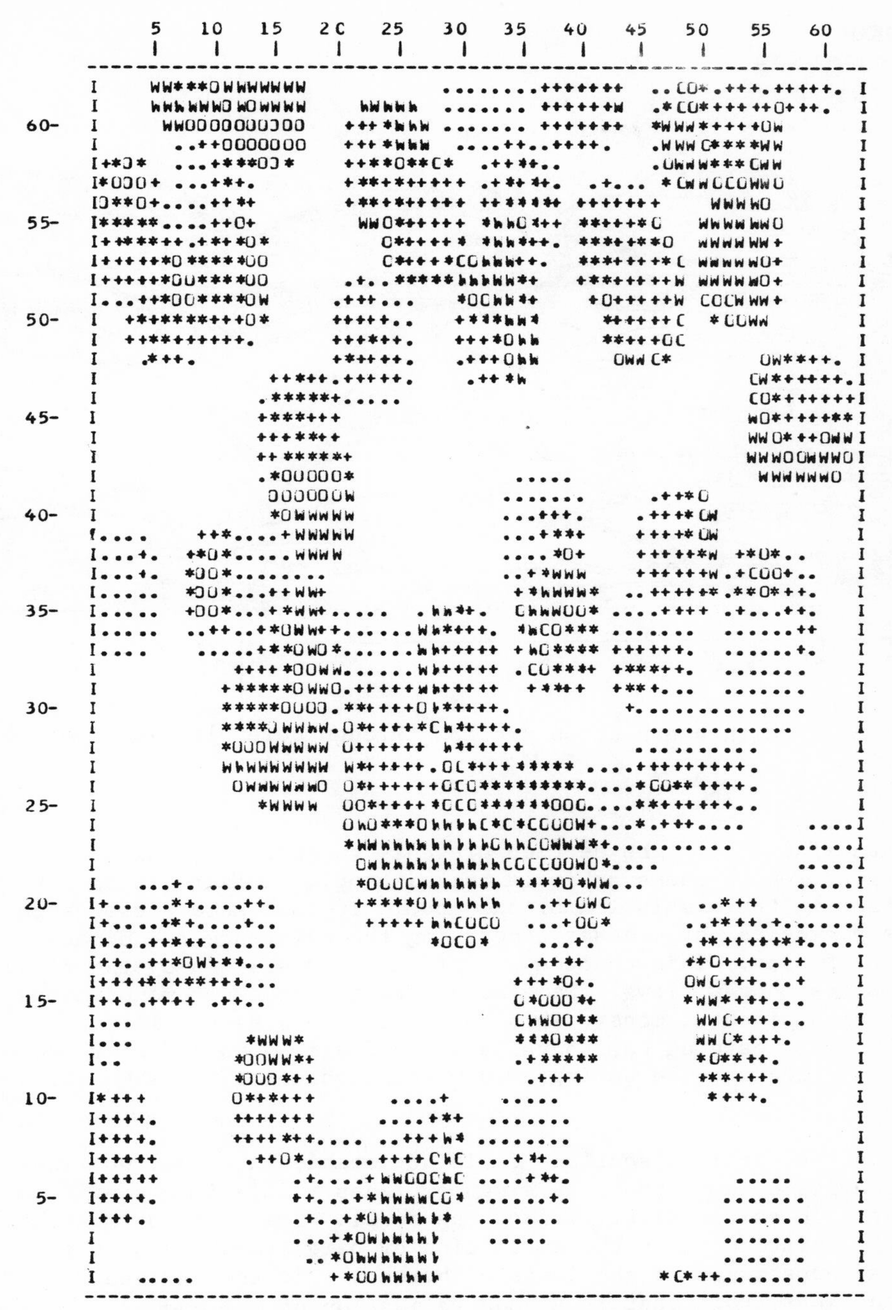

Figure 5. - Thickness map of layer No. 64 (clay layer). Scale: 1 unit=25 m (N-S direction); 1 unit=17 m (E-W direction). Empty space = blank; . = less than 0.4 m; + = 0.4 - 0.9 m; * = 0.9 - 1.5 m; 0 = 1.5 - 2.5 m; W = more than 2.5 m.

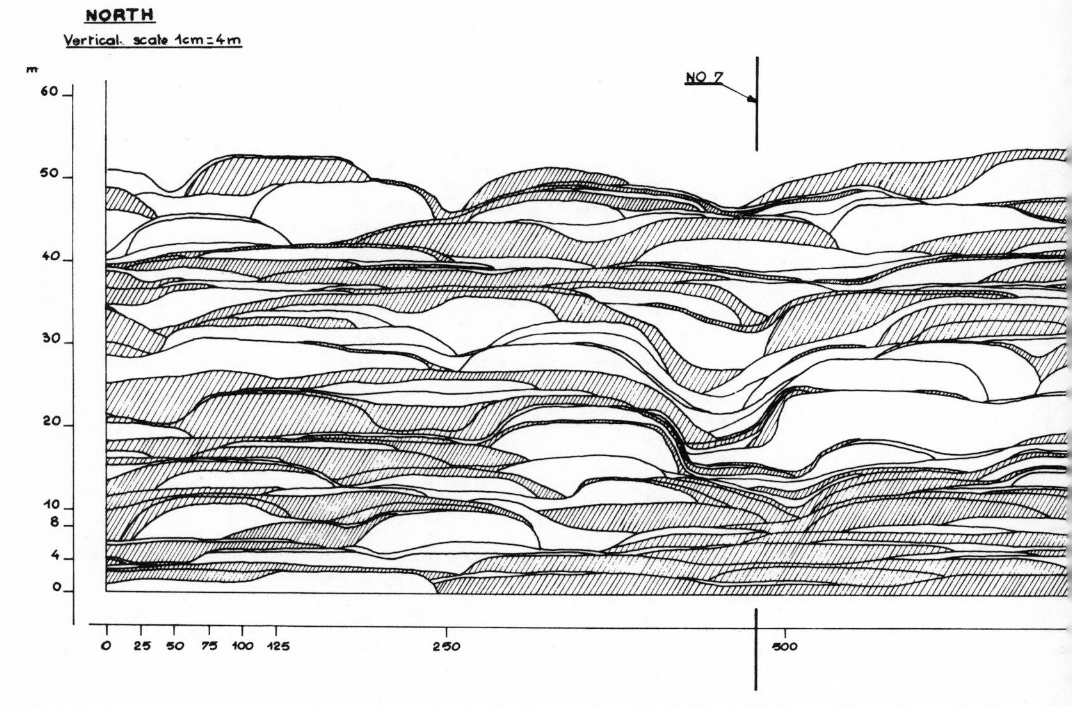

Figure 6. - North-south cross section through boreholes No. 7 and 8.

tal data (Table 1). The section presents an extremely chaotic appearance, which seems a little astonishing. In fact, it is not a mistake in the simulation or the model, it is a mere consequence of the geological hypotheses concerning the existence and dimensions of lenses. This chaotic appearance is sustained by the first experiments on gas flows. We also notice the somewhat monotoneous aspect of horizontal maps. This is due to the over-simplified hypotheses concerning horizontal shape and dimensions of microlenses. These hypotheses could be improved without changing the simulation method.

In the vertical section, it is noticeable that the two lines of boreholes do not appear as singular lines. This is a condition for the fitness and simulation of the model. Even a minor mistake in the simulation or in the estimation of parameters leads to a singular appearance of the lines of boreholes in the vertical section. Thus, this fact is a good validation of the model.

We must emphasize once more that as any simulation, these

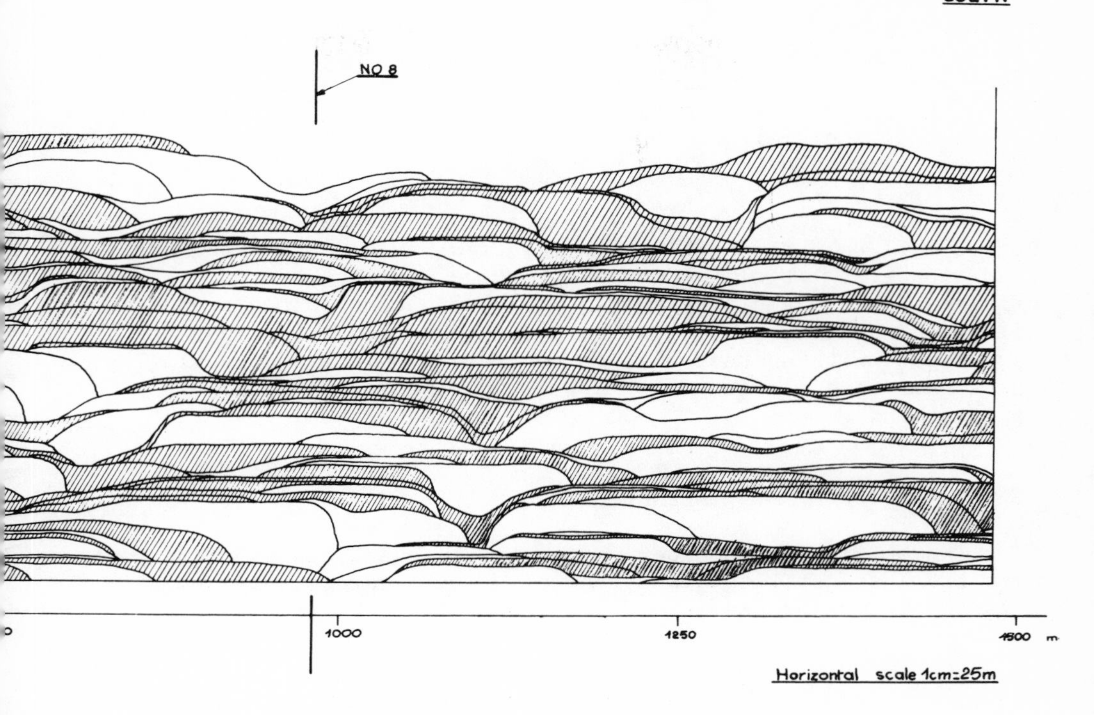

results give only a possible interpretation of the sedimentary structure. It is by no means an exact representation of conditions between boreholes.

CONCLUSIONS

When using certain random-genetic models, we know how to make a simulation of the sedimentary structure, conditionally to data collected in the boreholes. Moreover, the result is in an approximation of the most probable shape of the structure.

We have made several approximations and some of them have remained unjustified. The appearance of the results indicates that these approximations are fair, but it would be interesting to try a quantitative estimation of their range.

It is possible to generalize from this method to development of other random-genetic models. This scheme is specific to the model, especially the genetic hypotheses which we have assumed for the Chemery structure. For another model, even slightly different, it would be necessary to start from theory again. As this theoretical part is complex, it would represent a large amount of work. Moreover, the proposed solution is based upon an important hypothesis, the smallness of microlenses. Thus, for another model, we may not be able to solve the problem.

We finish on a more optimistic aspect. Indeed, we have obtained more than a conditional simulation; we have obtained the conditional law itself. Consequently, we may hope to solve other problems. For example we should be able to compute the probability, conditionally to available data, for one sandy lens to communicate with another one, or to treat problems of covering. These questions are far more interesting than a simulation in application.

ACKNOWLEDGMENTS

The work for this paper was partially supported by a contract with the Institut Francais du Petrole. Data were kindly communicated by Gaz de France.

REFERENCES

Feller, W., 1966, An introduction to probability theory and its applications, v. 2: John Wiley & Sons, New York, 509 p.

Jacod, J., and Joathon, P., 1970, Estimation des parametres dans les modeles aleatoires-genetiques representant des processus de sedimentation: Revue L'Inst. Fr. des Petr., v. 25, no. 10, p. 1144-1162.

Jacod, J., and Joathon, P., 1971a, Use of random-genetic models in the study of sedimentary processes: Jour. Intern. Assoc. Math. Geology, v. 3, no. 3, p. 265-279.

Jacod, J., and Joathon, P., 1971b, Rapport d'Etude sur la structure de Chemery: Int. report, Centre de Morph. Math.

Krumbein, W. C., 1967, FORTRAN IV computer program for Markov chain experiments in geology. Kansas Geol. Survey Computer Contr. 13, 38 p.

Krumbein, W. C., 1968, FORTRAN IV computer program for simulation of transgression and regression with continuous-time Markov models: Kansas Geol. Survey Computer Contr. 26, 38 p.

Krumbein, W. C., and Dacey, M. F., 1969, Markov chains and embedded Markov chains in geology: Jour. Intern. Assoc. Math. Geol., v. 1, no. 1, p. 79-96.

Schwarzacher, W., 1969, The use of Markov chains in the study of sedimentary cycles: Jour. Intern. Assoc. Math. Geology, v. 1, no. 1, p. 17-41.

APPENDIX

(1) Actual configuration. With the same notations as in the section on Research for Conditional Laws, we set,

$$q_j = m_j + n_j$$
$$q_1 = 1 + n_1 \qquad \text{if } n_1 = 0$$
$$q_A = m_A + 1 \qquad \text{if } n_A = 0$$
$$p_j = \sum_{i=j}^{A} q_i .$$

We need the following numbers (C_p^q is the combination of families of q objects among p)

$$F(Q, 1, (m, n)) = \begin{cases} C_{Q-1}^{m-1} & \text{if } n = 0 \\ \sum_{h=m-1}^{Q-n} C_h^{m-1} C_{Q-h-1}^{n-1} & \text{if not} \end{cases}$$

$$G(Q, 1, (m, n)) = \begin{cases} C_Q^m & \text{if } n = 0 \\ C_Q^n & \text{if } m = 0 \\ \sum_{h=m}^{Q-n+1} C_h^m C_{Q-h}^{n-1} & \text{if not} \end{cases}$$

$$F(Q,A,(m_i,n_i,)_{1\leq i\leq A}) = \sum_{h=q_1-1}^{A+Q-1-p_2} F(h,1(m_1,n_1))F(Q-h,A-1,(m_i,n_i)_{2\leq i\leq A}).$$

$$G(Q,A,(m_i,n_i)_{1\leq i\leq A}) = \sum_{h=q_1-1}^{Q+A-1-p_2} G(h,1,(m_1,n_1))F(Q-h,A-1,(m_i,n_i)_{2\leq 1\leq A}).$$

Now we can give the law of D_i. At first D_1 takes the values between q_1-1 and $Q+A-1-p_2$. The probability of $(D_1 = r)$, conditionally to $(m_i,n_i)_{1\leq i\leq A}$, is normed and proportional to

$$G(r,1(m_1,n_1))F(Q-r,A-1,(m_i,n_i)_{2\leq i\leq A}). \tag{11}$$

If we know $D_1,\ldots,D_{j-1}$, the number D_j takes the values between q_j-1 and $Q + A - j - p_{j+1} - \sum_{h=1}^{j-1} D_h$. The probability of $(D_j = r)$ is proportional to

$$F(r,1,(m_j,n_j))F(Q-r - \sum_{h=1}^{j-1} D_h, A-j,(m_i,n_i)_{j+1\leq i\leq A}). \tag{12}$$

Then we give the law of B_i. At first D_1 is known. If $m_1=0$, B_1 takes the values between 1 and $D_1 + 1 - n_1$, and the probability of $(B_1 = r)$ is proportional to

$$C_{D_1-r}^{n_1-1} \tag{13}$$

If $m_1 \geq 1$, B_1 takes the values between m_1 and $D_1 + 1-n_1$, and the previous probability is proportional to

$$C_r^{m_1} \quad C_{D_1-r}^{n_1-1} \tag{14}$$

B_i takes the values between m_i and $D_i+1 - n_i$. The probability of $(B_1 = r)$ is proportional to

$$C_{r-1}^{m_i-1} \quad C_{D_i-r}^{n_i-1} \tag{15}$$

If $i = A$ and $n_A \geq 1$, we still have (14). If $i = A$ and $n_A = 0$, then $B_A = D_A$.

(2) Lithological correlations. We assume that Q is known. So we have different possibilities to fix the number of each correlation layer. Each of these possibilities divide each borehole into several zones of known length. Let us suppose that drilling i is divided into N^i zones. The length of the kth zone is $Q^{i,k}$, and we have $\sum_{(k)} Q^{i,k} = Q$. This k^{th} zone consists of $A^{i,k}$ couples of macrolenses, containing $m_j^{i,k}$ and $n_j^{i,k}$ lenses. Then the probability for obtaining a certain system of numbers of correlation layers,

which gives the values $Q^{i,k}$, is proportional to

$$\prod_{(i)} [G(Q^{i,1},A^{i,1},(m_j^{i,1},n_j^{i,1})_{1\leq j\leq A^{i,1}}) \prod_{k=2}^{N^i} F(Q^{i,k},A^{i,k}(m_j^{i,k},n_j^{i,k})_{1\leq j\leq A^{i,k}})] .$$

The maximum likelihood method consists of taking the values of the numbers of the correlation layers which lead to the largest values, for the previous expression. For each borehole, the laws of $D_j^{i,1}$ and $B_j^{i,1}$ are given in the previous section. When $k \geq 2$, we must replace (11) by (12) and (14) by (15) to obtain the laws of $D_j^{i,k}$ and $B_j^{i,k}$.

(3) Law of S_n. The function $F(y)$ being defined by equation (4), we set

$$H(y) = 1 - \beta + \beta F[-\operatorname{Log}(1-y)] ,$$

and we denote the derivative of $H(y)$ by $h(y)$. Then we determine that the law of S_n conditionally to Y_{n-1}^{i+} and U_n^i admits a density. This density is normed and proportional to

$$e^{-\gamma t} \prod_{(i)} \left[\frac{1}{Y_{n-1}^{i+} + t} \; h\left(\frac{U_n^i}{Y_{n-1}^{i+} + t} \right) \right] .$$

(4) Layer in the influence zone of a borehole. We consider the influence zone of borehole i, that is an ellipse centered on this drilling. We know ΔX_n^i. If $\Delta X_n^i = 0$, there is no microlens centered in this zone. Let us suppose that $X_n^i > 0$.

A suitable affinity allows us to consider that the zone is a circle of known radius a. The function $N(z)$ being defined in section on Probability Laws we set for $\rho \leq a$,

$$K_\rho(y) = N\left(\frac{ya^2}{a^2 - \rho^2} \right)$$

$$\theta(y) = 2 \pi \theta \int_0^a \rho K\rho(y) \, d\rho ,$$

where θ is the Poisson parameter of the layer (θ_S or θ_C according to the lithology). As $X_n^i > 0$, there is at least one microlens centered in the zone. In fact, there is $P+1$ such microlenses, and

P is a Poisson random variable with parameter $\theta(\Delta X_n^i)$. The arguments of the centers of these microlenses are mutually independent random variables, uniformly distributed $(0,2\pi)$. The distances between the borehole i and these centers are also mutually independent random variables. For the P first microlenses, these variables admit density which is proportional to (the argument is ρ, and $\rho \leq a$),

$$K_\rho \; (\Delta X_n^i).$$

The law of their thickness is $N(z)$, properly truncated so that along the line of boreholes, the thickness is smaller than ΔX_n^i.

At least, let $k_\rho(y)$ be the y-derivative of $K_\rho(y)$. The distance between the borehole i and the center of the last microlens admits a density, which is proportional to (the argument is ρ, and $\rho \leq a$),

$$\rho \;\; k_\rho \;\; (\Delta X_n^i).$$

Its height is such that the thickness is exactly ΔX_n^i along the borehole.

AREAL VARIATION AND STATISTICAL CORRELATION

William C. Krumbein

Northwestern University

ABSTRACT

Correlations introduced between pairs of otherwise statistically independent and uncorrelated variates by similarity in their areal map patterns can be evaluated by partitioning the observed variance-covariance matrices into a trend component and a residual. If the original variates are correlated and have areal trends as well, the problem becomes complicated for sedimentological studies in which only one sample can be taken at any given map point. The situation becomes even more complicated when initially open data are closed, as by taking percentages.

This paper reviews the topic in terms of as yet unpublished work in the past two years. Several possible sources of covariance in sedimentary data are discussed, and it is emphasized that although the effect of map trends can be satisfactorily estimated, some other covariance components are confounded with residual effects. The general conclusion is reached that sedimentological correlation studies of environmental areas need to be supplemented by trend analysis to detect situations in which a seeming correlation is simply the result of similar areal responses by a given pair of variates to a common set of environmental conditions.

INTRODUCTION

Observational data in the earth sciences are especially characterized by two prominent features: (1) a tendency for phenomena to display systematic changes through time or in space, and (2) for variates to be complexly interlocked in such manner that correlations between them are the rule rather than the exception. As a result, maps that show areal variations and scatter diagrams that show one variate plotted against another, are commonly used for description, analysis, and interpretation of earth phenomena. The scatter diagrams are increasingly supplemented by regression lines and correlation matrices, especially since the computer became widely available. Similarly, the computer has made it relatively easy to fit least-squares surfaces to the map data, thus bringing out the systematic trends as well as local anomalies or residuals in the observed map patterns.

Until relatively recently these two approaches to data analysis have been made more or less independently of each other. In correlation analysis the data may be collected from different areas and combined into one set, whereas map analysis requires a set of observations in a specific area. As an example, the correlation between beach slope and mean grain size can be examined by taking one or a few samples from many beaches at different seasons, thereby avoiding the limitations imposed by sampling one beach or confining the samples to some single season. By the same token, if samples are taken from a single area essentially simultaneously, interest focuses on maps of individual variates, and correlations between the mapped variates may be only incidental to the main study of each map as an entity.

Complexities enter the picture when correlation studies are made on sets of samples collected from a particular area. In these circumstances the map patterns of any two variates may strongly affect the observed correlation between them. Thus, if any two variates are measured at the same sampling points in the area, and if they show similar map patterns, the observed correlation will be high, regardless of the nature of the variates themselves. It is entirely possible for two variates that have no real or inherent correlation between them to show a high seeming correlation simply because their areal patterns of variation are similar.

This correlation arises because similar maps have large values of the mapped variates in the same part of the map area, and smaller ones in the down-trend direction. If the patterns are such that one set of values increases as the other decreases, the correlation will be large, but negative. It is only when the two maps have contours essentially normal to each other that the map patterns do not induce a correlation between the pair of variates involved. What this implies is that correlation coefficients computed from the observed values at common sampling points in the same map area need to be examined somewhat more closely than heretofore.

A SIMPLE PARTITIONING MODEL

For some years it has been apparent at least intuitively that some relation is present between areal patterns and correlations. In 1960 Mirchink and Bukhartsev used the Pearson correlation coefficient r to express the similarity between structure maps drawn on successive subsurface horizons. This earlier work, plus some seemingly aberrant scatter diagrams of sedimentary properties, induced Jones and me (Krumbein and Jones, 1970) to look into the question more formally. We were able to show theoretically that areal trends introduce a covariance component between two mapped variates entirely independently of the nature of the variates being mapped; and that this covariance can be exactly predicted from the trend-surface coefficients and the spatial arrangement of the map control points.

Our first model, expressed at the time as

$$SS(X,Y)_{\text{Total}} = SS(X,Y)_{\text{Trend}} + SS(X,Y)_{\text{Residual}} , \qquad (1)$$

stated that for two independent mapped variables X and Y not themselves "inherently" correlated, the total corrected crossproduct sum of squares can be partitioned into one part contributed wholly by the map trends, plus a nonsignificant residual covariance arising from sampling error and other random effects. The model is somewhat more convenient if covariances are used instead of sums of squares. The results are the same, but relations among several pairs of mapped variates can be more compactly shown by variance-covariance matrices and their accompanying r-matrices.

Studies made since 1970, but not yet published, showed that if the two mapped variates are inherently correlated and also display trends, the trend component can be isolated, but the inherent covariance is assigned to the residual, and hence is confounded with the random effects. If the error terms are small, the residual is a fair estimate of the inherent correlation.

More recently (Krumbein, 1972) the problem of correlations introduced when a set of open variates with trends is closed (as in converting rock thicknesses into percentages) was investigated. The definitive work in this subject is by Chayes (1960, 1971) who developed the theory for covariances introduced by closure of open systems without trends or inherent correlations. If trends are present in the open system, the situation becomes complicated on closure, and this complexity increases if both trends and inherent correlations occur in the open system. My 1972 paper was expository, and the problem is now being looked into theoretically (Krumbein and Watson, 1972, in preparation).

The purpose of this paper is to discuss the implications of the model in equation (1) in terms of the need for distinguishing between several sources of covariance in sedimentological data. My thesis is that full interpretation of observed correlation coefficients requires simultaneous study of map patterns and correlation in those problems that involve areal variation.

SOURCES OF COVARIANCE IN SEDIMENTOLOGICAL DATA

The term inherent correlation is somewhat vague. A better term for what is intended is "correlation at a point", which refers to correlations between pairs of open variates independently of any areal or temporal connotation, such as the correlation between beach slope and mean grain size mentioned earlier. Trend correlation is induced between two mapped variates solely by the systematic areal patterns or trends that they display. Closure correlation arises entirely from closure of an open system. In its simplest form this represents the closure of an open system without point correlations or trend correlation. If these are present, the situation becomes complicated, as mentioned. A fourth source of covariance in observational data is the "accidental" correlation, usually statistically nonsignificant, that may occur even if pairs of numbers are drawn at random from two uncorrelated populations without trend. These are generally statistically nonsignificant, and the term error correlation seems appropriate.

Conceptually, the simple model of equation (1) can be extended to include these several sources of covariance, as follows

$$\mathrm{Cov}_{\mathrm{Total}} = \mathrm{Cov}_{\mathrm{Trend}} + \mathrm{Cov}_{\mathrm{Point}} + \mathrm{Cov}_{\mathrm{Closure}} + \mathrm{Cov}_{\mathrm{Error}}. \quad (2)$$

Some constraints on geological sampling, plus the present state of knowledge, place severe limitations on our ability to partition the total observed covariance into all of these specific categories, however.

PROBLEMS IN COVARIANCE PARTITIONING AND MODEL TESTING

Given two maps with open variates Z_1 and Z_2 measured at the same control points, the total covariance is computed directly from the pairs of observed values. As mentioned, the trend component can be exactly predicted.

Covariance at a point cannot be estimated unless replicate samples can be taken at each map control point. This can seldom be done in geology inasmuch as collection of a rock specimen leaves a void. For most problems we are limited to one sample at any one point. Clusters of closely spaced samples at each map point may approximate replicate sampling, but this requires additional study. Similarly, covariances due to closure present difficult estimation problems in data initially closed, such as heavy-mineral percentages or chemical composition of sediments. When open data are deliberately closed, as in apportioning particle size into percentages of sand, silt, and clay, strong closure correlations are introduced, but in some instances these are predictable from the closed variances (Chayes, 1960; Krumbein and Watson, 1972, in preparation).

Additional problems rise when real-world data are used to test the models of equations (1) and (2). In the long run this is obviously necessary, but in the present state of knowledge one cannot fully evaluate the results of a correlation-trend study with field data. This is because none of the population parameters, including the covariance components, is known. Improvement in methodology requires Monte Carlo experiments based on large random samples from fictitious populations with assigned parameters. This has been my main approach, although parallel studies are being conducted on beach foreshores that hold promise of yielding meaningful results. A third approach is to use the results of real-world studies and to work backwards, by iterative Monte Carlo methods, to find combinations of population parameters that give the same results as those observed in nature.

As stated earlier, we found variance-covariance matrices more directly useful than sums of squares and crossproducts. As an illustration the gamma-distributed data used in Jones' and my 1970 paper is resummarized here in Table 1 to illustrate the present form of our output. For two mapped variates X and Y the matrix elements are 11 = Var X, 22 = Var Y, and 12 = 21 = Cov(X,Y), but the top matrix, with the total observed map data, is additively partitioned into the two lower matrices, showing the trend and residual components respectively. On the right are the corresponding r-matrices.

Our first experiments were somewhat cumbersome because we did not assign simple integer or fractional values to the fictitious population parameters. Even so, the matrices on the left of Table 1 show clearly that with respect to the residual matrix, the much inflated values in the top left matrix have been contributed by the trend components in the center matrix. The trend maps (Jones and Krumbein, 1970, figs. 5 and 6) have contours so nearly parallel that the trend r is essentially 1.00.

Table 1. - Variance-covariance matrices for gamma-distributed data from Krumbein and Jones (1970)*

"Observed" Map Data

$$\mathrm{Cov}_{\mathrm{Total}} = \begin{bmatrix} 0.06102 & 0.04021 \\ 0.04021 & 0.02705 \end{bmatrix} \qquad r_{\mathrm{Total}} = \begin{bmatrix} 1.000 & 0.989 \\ 0.989 & 1.000 \end{bmatrix}$$

Trend Component

$$\mathrm{Cov}_{\mathrm{Trend}} = \begin{bmatrix} 0.06011 & 0.03952 \\ 0.03952 & 0.02600 \end{bmatrix} \qquad r_{\mathrm{Trend}} = \begin{bmatrix} 1.000 & 0.999 \\ 0.999 & 1.000 \end{bmatrix}$$

Residual Component

$$\mathrm{Cov}_{\mathrm{Resid}} = \begin{bmatrix} 0.00090 & 0.00068 \\ 0.00068 & 0.00105 \end{bmatrix} \qquad r_{\mathrm{Resid}} = \begin{bmatrix} 1.000 & 0.716 \\ 0.716 & 1.000 \end{bmatrix}$$

*See Tables 5 and 6 of cited reference for sums of squares analysis. Sample size, n 20.

The residual matrix itself approaches closely to the values of the nontrend population parameters, in which Var X = Var Y, and the covariance agrees exactly with that in the gamma distribution used for the experiment. The residual r of 0.716 also is close to its expected value of 0.70. Thus by arranging our original output in the form of Table 1, it seems fairly certain that the point correlation between the mean and standard deviation of the gamma distribution is assigned to the residual rather than to the trend. Numerous subsequent Monte Carlo runs with a bivariate normal distribution having known trends and known point correlations amply demonstrate that this allocation occurs.

In terms of the expanded model of equation (2), then, present methodology for studies in which only one sample can be collected at a given map point, we cannot isolate the point covariance from the residual. Whether the closure covariance can be isolated in open systems that are subsequently closed is now being examined; for data initially closed the situation looks somewhat less hopeful.

CONCLUDING REMARKS

In essence this paper is an interim report on a problem that arises in many sedimentological studies. Where the researcher can be confident that no point correlations are present, the simple model of equation (1) satisfactorily removes the trend effect from correlation analysis. If a point correlation is present, the trend component can be removed, and the residual can be used as an approximate estimate of the point correlation free of areal effects. As long as open data are involved, the closure covariance is zero and causes no concern.

The search for correlations among sedimentary properties is an integral part of sedimentological analysis. Moreover, the r-matrix is the starting point for factor analysis, cluster analysis, stepwise regression, and other applications of the general linear model. Whether it is important to separate the trend effect from point correlations in these applications may depend on the purpose of an individual study. Nevertheless, knowledge of the structure of his r-matrices gives the researcher a choice of techniques not available to him if he does not also examine the areal implications of his observational data.

REFERENCES

Chayes, F., 1960, On correlation between variables of constant sum: Jour. Geophysical Res., v. 65, no. 12, p. 4185-4193.

Chayes, F., 1971, Ratio correlation: Univ. Chicago Press, Chicago, 99 p.

Krumbein, W. C., 1972, Areal variation and statistical correlation in open and closed number systems: Intern. Assoc. Stat., Phys. Sci., Proc. (in press).

Krumbein, W. C., and Jones, T. A., 1970, The influence of areal trends on correlation of sedimentary attributes: Jour. Sed. Pet., v. 40, no. 2, p. 656-665.

Krumbein, W. C., and Watson, G. S., 1972, The effect of trends on correlation in open and closed three-component systems: Jour. Intern. Assoc. Math. Geol. (in preparation).

Mirchink, M. F., and Bukhartsev, V. P., 1960, The possibility of a statistical study of structural correlations: Dokl. Earth Sci. Sect., v. 126, no. 5, p. 1062-1065 (English Translation, p. 495-497).

FORMATION AND MIGRATION OF SAND DUNES: A SIMULATION OF THEIR EFFECT IN THE SEDIMENTARY ENVIRONMENT

Michael J. McCullagh[1], Norman E. Hardy[2], and William O. Lockman[2]

[1]Kansas Geological Survey and [2]The University of Kansas

ABSTRACT

A computer-simulation model has been constructed to study the effects of changing wind strength, direction, and supply on the formation and migration of various types of dunes. The track of the dune through time is studied by means of a graphical map output system that draws maps of the dune's development at desired intervals. The changes in form and in rates of erosion and deposition then can be studied in detail.

INTRODUCTION

Only a limited amount of work has been done on the formation and migration of sand dunes and their effect in the sedimentary environment. Much early work was carried out by Bagnold (1937, 1951, 1954) and appeared in the days before high-speed computers. Allen (1968) has collated and developed the theory of particle movement in water, but there is little work directly related to sand flow in air, or to the exact formation of macrofeatures. The purpose of this paper is to examine the existing body of theory relating to sand dunes and using this knowledge, construct a computer simulation to observe the effects of changing wind speed, direction, and sand supply on the formation and migration of various types of dunes.

The computer simulation, developed on the Honeywell 635 at The University of Kansas, produces a two-dimensional plot of dune development at desired time intervals. Wind strength, direction, and sand supply, along with other parameters, can be altered to fit any desired combination from either a hypothetical or actual example. Any profile may be chosen for a beginning. Then the predetermined conditions acting on this surface simulate movement of the sand and, in the long run, entire dunes. Output, produced on an online plotter, is in the form of a profile or a series of profiles at desired simulated time intervals. By changing the probability of the controlling conditions, empirical tests can be made of the theoretical treatment put forward by various authors dealing with the migration and stabilization characteristics of sand dunes.

First, we will review the existing theory on the subject. This section will be followed by one treating the program and its development, and finally, a section discussing examples produced by the program.

THEORY OF SAND MOVEMENT

Sand dunes, like deltas, flood plains, glacial deposits, and other sedimentary features, depend upon a positive balance of an appropriate material within a given area for their formation. Thus, where wind moves over a sand source into an area in which its velocity decreases, a positive sand balance will exist and the result is some type of dune field.

As Bagnold (1954, p. 175) notes, these conditions may occur if the direction of storm winds differs from the prevailing direction of the normal, more gentle winds. Here, the storm winds bring new material whereas the prevailing gentle winds provide a continuous shaping effect. As prevailing winds continue, sand from the windward side of the dune is moved over the top and around the edges of the dune, and dune migration is begun. Three requirements are necessary for this process to continue: (1) the sand source must continue to exist, (2) storms must occur occasionally, and (3) prevailing winds must continue.

Allen (1968) includes a section dealing with sand dunes and their remarkable similarity to water ripples. He notes that without disturbances, dunes tend to maintain a constant form. Also, they must be reasonably well spaced or isolated so that one dune does not interfere with another. In modeling it is necessary to ensure that dunes maintain a nearly constant shape for as long as controlling conditions are similar. If the spacing between dunes becomes too small, their development and migration must be affected.

Direct measurements of the rate of movement of sand per unit time have been made by Bagnold (1954, p. 69-70) both in the desert and in wind-tunnel experiments. The quantity of sand moved increases exponentially with wind velocity, thus

$$q = 5.2 \times 10^{-4} \ (v - Vt)^3 , \qquad (1)$$

where q = quantity of sand movement in metric tons per hour for a lane 1 m wide ,

v = wind velocity in m per sec, and

Vt = threshold or critical wind speed in m per sec above which sand is in motion.

This relationship is used in the simulation and is illustrated graphically in Figure 1. The critical wind speed varies according to the size of particle being transported and can be calculated approximately by equations relating drag effect, surface roughness, and wind velocity at a given height above the sand surface.

No fully satisfactory theory is available for the distance that sand from a point source will be moved for a given windspeed

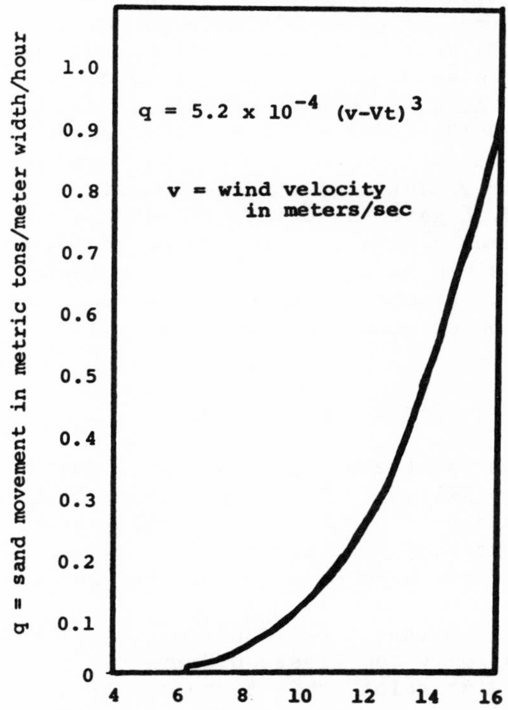

Figure 1. - Relationship between quantity of sand movement and wind velocity at standard height of 1 meter (after Bagnold, 1954, p. 70)

or for the distribution of its deposition. It has been assumed in this study that the downwind distribution from a point will approximate a negative exponential function leading to decreasing deposition with increasing distance. The result of this hypothesis is that movement is totally controlled by windspeed and gradient.

Wind velocity near the ground on the leeward side of a dune is decreased owing to the protection offered by the dune itself. This is illustrated in Figure 2. Thus, the lee area of a dune becomes an area of potential deposition. Sand from the dune's windward side is moved over or around the crest and deposited on the leeward side. If this slope is less than the angle of repose for the material, about 34° for most sand dunes, sand will be deposited; alternatively, if the slope is equal to the angle of repose, sand will tend to move down the slope to collect at the bottom.

Alterations in the windspeed and local gradient of the sand surface lead to changes in the shape of the dune caused by differing sand deposition from what can be considered to be an infinite number of point sources. If no local gradient exists, windspeed is constant over the relatively small area in which an incipient dune might form. As a simplifying assumption, the dunes in this model possess only one sand size. Thus, as movement would be uniform over such a surface, no dune creation can take place. Both Allen (1968) and Bagnold (1954) point out that dunes can form only where more than one particle size is present, as formation is due to the differential movement of the different sizes of particles if exposed to a wind of given velocity. This does not preclude either the migration of a dune or a change of shape. Neither is a growth in size impossible as long as there is a sand supply that exceeds removal. This situation can occur if sand supply is at an angle to the direction of dune forming winds.

SIMULATION SYSTEM

The main problem in attempting to simulate dune change and migration is the complexity of the relationships involved and the large quantity of data that has to be available at any instant to determine how the next step will proceed. To circumvent this problem, simplifications have been made to the real-world situation.

First, and most important, the sand-size distribution is considered to consist of only one class, 0.3 mm, the median size found in many dunes. The distortions this feature introduces into the model are two. The fact that no dunes can be created from a flat plain entails that the simulation must use a measured dune profile as input. This profile then is modified and possibly migrated by the simulation program. Second, the change in shape of a dune is partially dependent on the size distribution of the sand at every point on the surface. This distribution changes with deposition and erosion. For a fully realistic model the recording of a continuous size distribution is essential, but unfortunately impractical. A certain degree of error therefore is brought into the changes wrought by the model on the input profile.

It is impossible, for similar reasons to those already outlined, to consider the profile of the dune as a continuous surface. For ease of computation the dune is divided into 150 cells of 1 m in length. The profile is one dimensional therefore a major problem is that of which profile to choose. As changes in the shape of the dune are most pronounced in the major wind direction, this direction is taken to be the most suitable for the profile.

Sand movement is simulated along the profile, on the assumption that side movement effects cancel out for any particular time period when a dune modifying wind is blowing. Entry of sand from the direction of any sand source that may be present is going to be critical in enlarging the dune system. The effect of the source on the profile will depend on the relation of the source winds to the profile directions. These factors are taken into account in the model.

The last major restriction employed in the model is related to end effects brought about because the scale of the model is not infinite. Figure 2 is a profile of two dunes, and implicitly assumes that there is some type of dune or sand field behind the last dune and in front of the first one. Translating this implicit statement into a simulation model is not easy as at some stage a boundary must exist. The solution used has been to assume that the selected profile is only part of a greater dune field, thus making conditions before and after the profile similar. It does not seem unreasonable, as the profile is part of the dune system, to consider the loss of sand from the right end of Figure 2 to be input

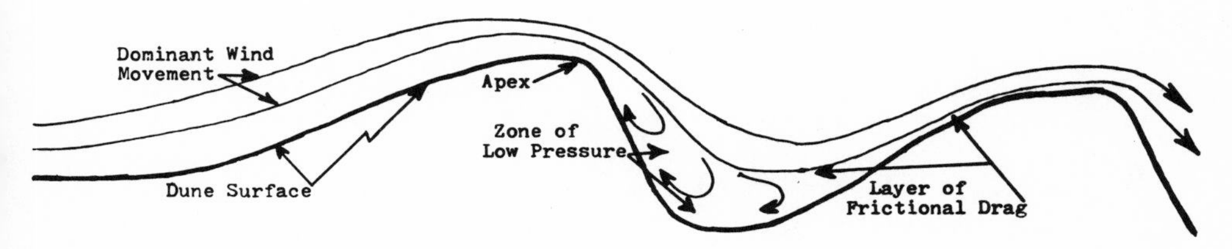

Figure 2. - Typical dune system.

sand to the left end of the same figure, thus forming an endless loop. Sample outputs from the model will show how although the sand travel forms an endless loop, conditions at the beginning of the loop can be changed to simulate the effect of the wind passing over different types of dune fields before the profile is reached. In other words, the "tail-eating" aspects of the model mean that the system is closed only in relation to total sand content present in the profile that has not been introduced from the source direction.

PROGRAM OPERATION

The operation of the program entails three calculation phases: windspeed changes caused by lee effects, local sand gradient values, and sand movement based on the previous two. The simulation is split into time increments or iterations, and performs the three calculation sets for every iteration. One time increment is approximately one hour of effective wind - wind that is strong enough to move some of the sand on the dune. The wind speed is assumed to follow an exponential distribution with a few increments of strong winds and many weak ones. The program uses the mean of this distribution as input. Obviously the direction of the wind is important as the amount of sand moved in the profile direction is a resultant of the two directions involved and can be considered an alteration in the apparent wind speed in the profile direction. Thus the winds can blow either direction along the profile, their velocities being the resultant of direction and actual speed. If the wind in the profile direction is effective, it exceeds the threshold velocity on any part of the profile, then the three phases of calculation commence.

Figure 2 shows isovelocity lines passing over the dune sequence. Velocity at a given height drops in the lee of the dunes and an area of lower pressure, and calmer or still air is present. It is necessary for the calculation of the quantity of sand that will be moved to know the windspeed close to the sand surface. Therefore, some algorithm is needed to determine how far into the lee of a dune a particular area of sand surface may be and also how much effect this distance will have on the windspeed. On an open sand surface, the windspeed at a given height can be calculated by

$$v = 5.75\, V_* \log \frac{Z}{k}, \tag{2}$$

where V_* = drag velocity,

Z = height, and

k = a constant dependent on surface roughness equal to 0.01 for particles of .3 mm.

This equation will not hold in the lee of a dune where the decrease in velocity from the open air above the dune is roughly linear to the sand floor in the lee. Although the windspeed is determinable in the open air above the dune by equation (2), the effect of the lee is not and is calculated in the model by a linear function.

It is therefore necessary to determine the vertical depth of the sand surface in the lee below a given constant windspeed height. The algorithm proceeds from a starting height for constant windspeed and then considers every point on the dune in relation to it. Where, as in the situation of the first dune in Figure 2, there is no lee effect until the crest of the dune is reached, the calculated lee height is obviously zero. As computation passes down the slip face of the first dune the sand surface passes into a lee, only emerging again about two-thirds of the way up the side of the next dune. Where the first dune is larger than the second it is possible that the second will always be in the lee of the first, and will not change shape or move much in relation to its taller counterpart. This feature is exhibited in later illustrations of model results.

The second phase of the iteration is to calculate the gradient between all 150 adjacent points on the dune profile. When sand movement takes place, the amount deposited is modified by these values. The relationship cannot be considered linear or even the same for upslope and downslope gradients. If the gradient exceeds 34° (angle of rest) in the downslope situation no material would be deposited, whereas in the upslope case only heavy rather than complete deposition would occur. The steep upslope would not be a permanent barrier to the movement of sand over the crest of the dune. The variation in deposition owing to gradient is complex and best measured empirically in the field. In the model various distributions have been tested; the best results were obtained using a logarithmic function giving high deposition rates for high gradients upslope and nil deposition for the extreme downslope situation.

The last, and most time-consuming, phase is the calculation of the sand movement based on the previous two phases and the windspeed vector for the present iteration. As the time increment being used, 1 hour, is long, a simultaneous removal technique along the entire profile has been adopted coupled with a subsequent addition of the incremental deposition quantities to the individual elements of the profile. Time restrictions negated the possibility of using a continuous approach.

Using equation (1), the quantity of material that will be removed from a square meter (the incremental unit along the profile is calculated for an incremented position along the profile taking into account the wind speed, source contribution, and any reduction caused by a lee effect. This material then is distributed downwind according to the function described previously. As deposition occurs sequentially downwind, the effect of gradient modifies what will be deposited. Thus on an uphill stretch it is likely that the deposition of the entire load may have occurred before the distance calculated for final deposition. The distribution curve will have been squashed. Conversely, downhill, the curve may be stretched considerably by the effect of gradient. The amount of deposition that accumulates in every unit of the profile is put into temporary storage until the movement of sand for all points along the profile has been calculated. Once the sand movement for every point is determined, the negative and positive accumulations are added to the profile to give the profile input to the next iteration.

GRAPHICAL OUTPUT

The process of lee calculation, gradient determination, and sand movement is continued for as long as necessary to produce the results required. On command, it is possible to make the program draw the profile or store it for future reference. By storing every fifth or tenth profile and comparing the results movement and changes of shape can be detected.

Two types of final profile are used for display of output. The first type shows a series of profiles produced after a selected number of iterations. Each profile is drawn in entirety and superposed on the previous one. A second type of output permits the plotting of only the final profile and any parts of earlier profiles that may be present inside the final one. Thus, the result is the final cross section of the dune formation exhibiting all calculated bedding features. It is important to note that the width between profiles is indicative of the deposition rate at that time. On all outputs the vertical scale is exaggerated by a factor of three compared with the horizontal.

EXAMPLES OF OUTPUT

The figures illustrating this section are on-line plots of runs of the simulation program. By changing input parameters in the program an endless number of experimental profiles could be produced. In this section a group of profiles is examined where input parameters are changed while holding the starting profile constant. The initial or starting profile selected is one where former prevailing winds usually blew from left to right. In short, the right slope is slightly steeper than the left slope as if the

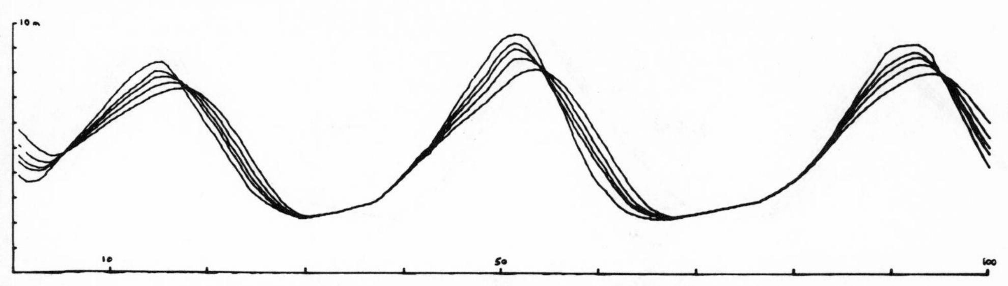

Figure 3. - Standard run; wind direction is from left to right.

right slope was a leeward slope.

Figure 3 shows the type of output produced by the program. The starting profile has gradually reduced in height and advanced to the right. The initial profile is relatively undunelike in that both sides are steep and there is little appearance of a predominant wind direction. The purpose behind this is to discover whether the simulation system is capable of turning an unlikely shape into a dune or dune chain. If it succeeds it means that the dune shape will, as it should be, be totally dependent on the wind gradient and grain size. The windspeed in this instance is set at roughly 5 m per sec resulting in a gradual decline in height, lengthening of the windward slope, and advancement and shortening of the lee slope. The simulation is working under the assumption, in most of these figures, that the profile section is in the midst of a dune field. Thus the first dune in the sequence is neither accreting or degrading more rapidly than the rest.

The remaining figures are all variants of this standard run and deal with particular problems and options of the system. Figures 4 and 5 show increases in source supply to the dune. In Figure 4, the supply from the sand source is moderate resulting in almost no change in dune height with only some deposition on the leeward size. This is because the source is traveling in the same direction as the predominant dune forming wind (left to right) and is replenishing the losses on the windward slope. A greater increase in supply is shown in Figure 5. The results show an increase in height and an increase in leeward deposition. The change in the windward slope is smaller. In the initial profile, the lee slope is the steepest and less deposition occurs; if the slope angle is less, the deposition is greater. Thus at first, deposition builds up at the bottom of the lee slope, but eventually the slope is reduced to the point where the net result is a parallel advance.

The average wind speed can be increased or decreased at will.

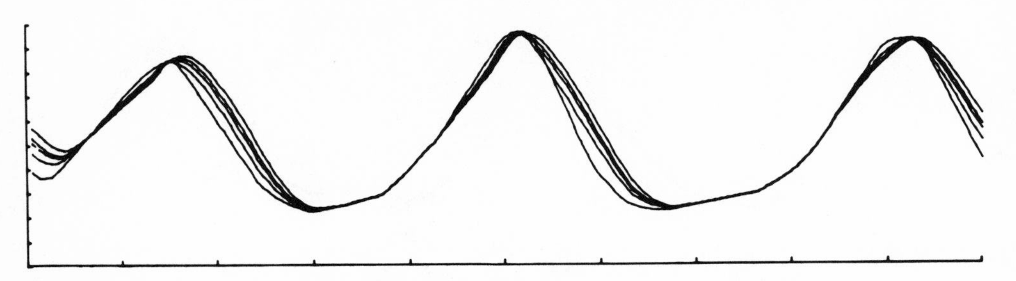

Figure 4. - Slight increase in source material results in constant dune height.

In Figures 6 and 7 the results of an increased average speed together with increased source supply is exhibited. For both examples the wind has a velocity of 6 m per sec instead of 5 as previously. Because rapid change is noticeable, the program was allowed to produce only four plotted cross sections before termination instead of five. Both examples reflect the increase of wind speed in greater migration and rapid decrease in height. Here, the force of the wind is strong enough to reduce dune height significantly in a relatively short period of time. In Figure 7, the sand supply is increased to the level of that in Figure 4 which results in a slow down of migration and a slower rate of change in height of the dune chain.

It may happen in practice that more than one wind direction is forming the dune feature. Established theory suggests that for seif dunes, alternate wind directions partly caused by dune changes farther up the chain may cause the interesting cross-bedding features present in such dunes. Figure 8 shows the result of a short test of this theory if the wind is allowed a 70% probability of blowing from the left of the profile. Because winds may come from either of two directions, dune migration is not as rapid. The result of deposition on both sides of the dune is apparent as is a slight decrease in height. Owing to the predominance of winds from the left, however, the lee slope associated with these winds is

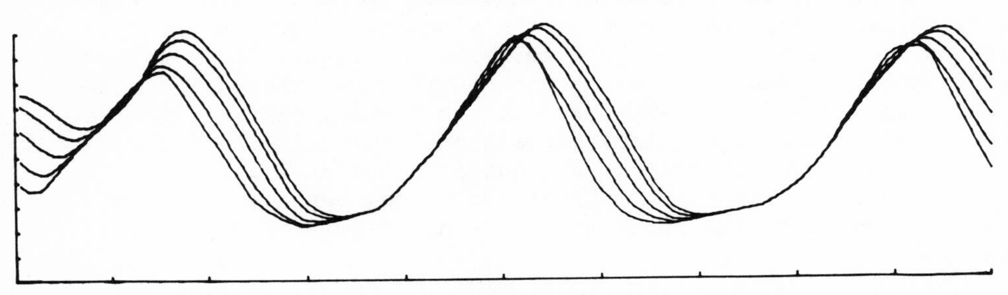

Figure 5. - Large increase in source material results in increased height.

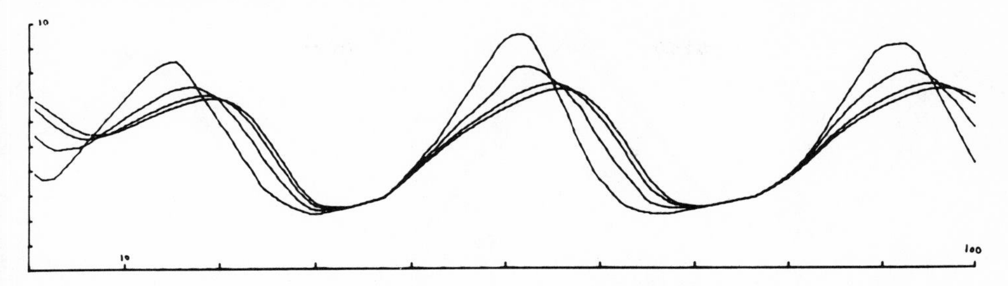

Figure 6. - Increase in wind speed.

the one to have advanced.

More than five superposed profiles tends to lead to confusion in the beholder. Therefore, for long simulation runs the cross section approach mentioned at the beginning of this section is adopted for the remainder of the figures. The advantage of this method is the lack of confusion as to which line is the final profile. The disadvantage is the loss of any part of the starting or intermediate profiles which were removed by subsequent erosion.

Figure 9 is the same as the standard run, Figure 3, except the program has been allowed to run for 400 significant intervals when the wind speed was above critical. Note that the deposition on the leeward side is apparent as well as considerable reduction in height. Also the dunes now have attained a stable and dunelike shape. Also note from the width between adjacent section lines that deposition has slowed although the average wind speed has been maintained throughout at 5 mph. The dips between the dunes have filled, but at the same time become more angular than they were at the start owing to the lee effect producing almost no wind movement in the trough.

In Figure 10, 270 hours of windspeed above critical occurred. The difference between this figure and the previous one is that a

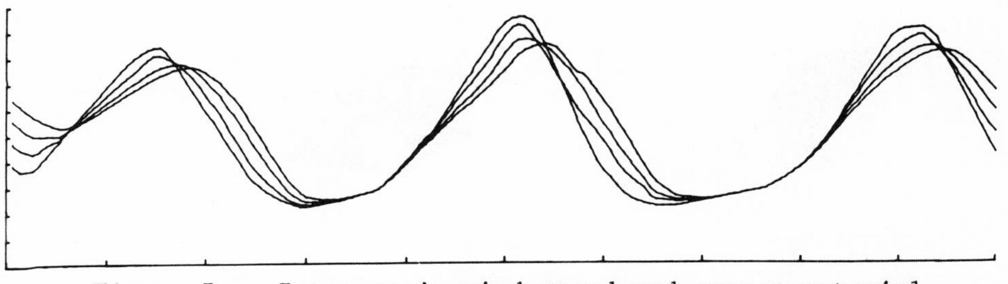

Figure 7. - Increase in wind speed and source material.

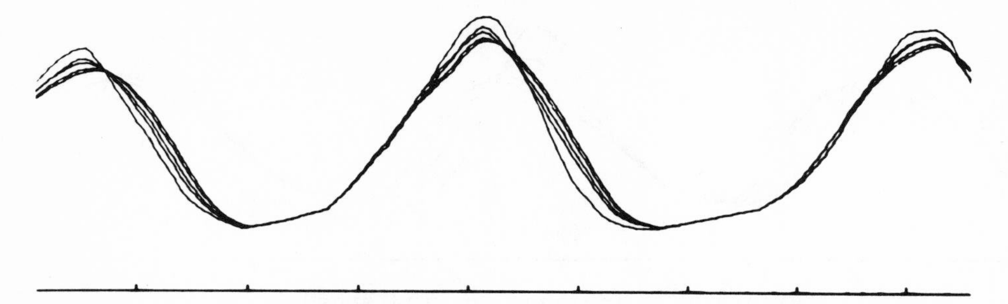

Figure 8. - Wind blowing from either left or right and slight increase in source material.

moderate growth factor was introduced. This led to extreme development of the left dune because the dune field outside the simulation area, not being updated as it only exists by inference, had not grown sufficiently to keep pace with the chosen profile. Thus the left dune has accreted steadily using this unchanging surrounding dune field as a source of supply.

Figure 11 illustrates the results over a long period of time for a mean wind speed of 5 m per sec from either direction, but a 70% probability of a forward wind (c.f., Fig. 7). After 400 significant intervals, height is decreased, numerous beds of deposition are found on what was usually the leeward side whereas most deposition on the usual windward side has been removed. Where both wind directions have an equal probability of occurring, a result as in Figure 12 is produced. After 100 significant hours the dunes have become dome shaped and exhibit accumulations on both sides. The lower diagram in the figure provides a comparison with the standard run.

Figure 13 shows a situation which occurs in reality. The probability of a forward wind is 70%, but the source direction is linked to the probability of a wind blowing backward (from right to left). Thus accretion can occur only if the wind is blowing from

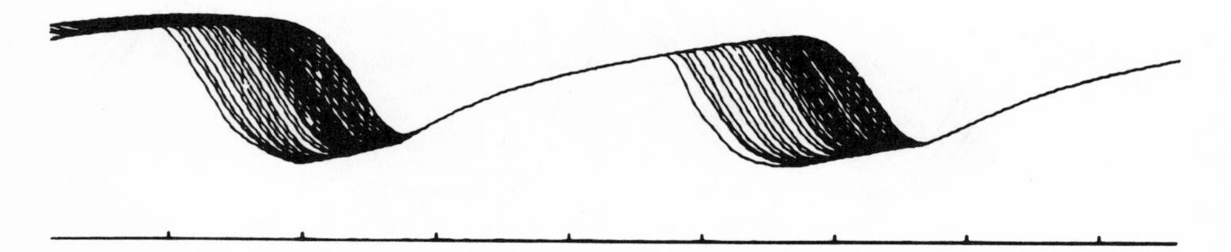

Figure 9. - Standard run after 400 significant intervals with only final profile and cross sections of deposition plotted.

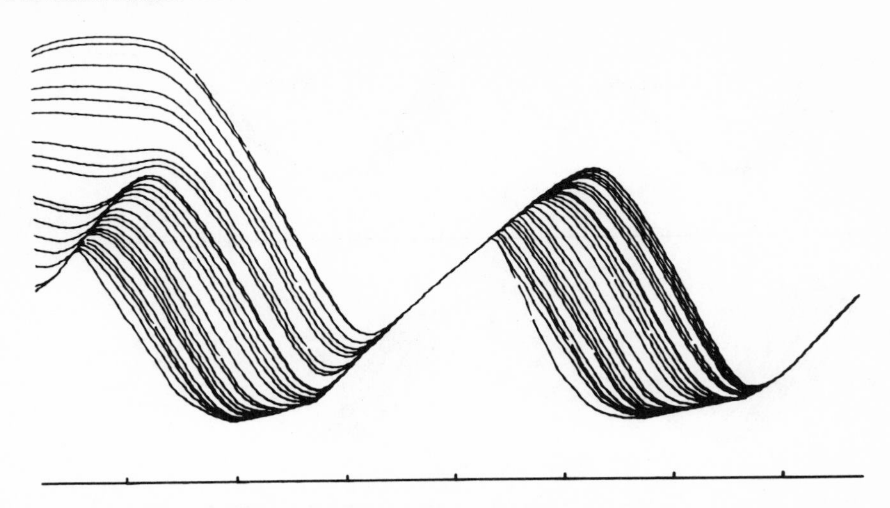

Figure 10. - Increase in source material after 270 significant intervals.

the right. The result is that the middle dune first of all grows and migrates to the right, and then once the left dune has formed a barrier putting the rest of the dune chain into its lee, continued growth is to the left, toward the left dune. Thus the dunes can start migrating toward the dune in whose lee they form, although this is opposite to the direction of the dominant dune-forming wind. If there is an equal chance of the wind blowing in either direction the resulting accretion is mixed on both sides of the dune system, as in Figure 14. The results show greater deposition on the right side than on the left side due to the original steeper right slope in the starting profile.

CONCLUSIONS

The advantages of a computer simulation over a scale model approach are primarily the ability to consider many variables,

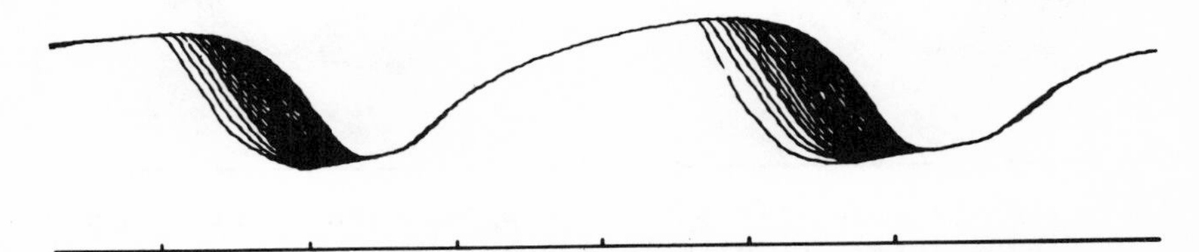

Figure 11. - After 400 significant intervals with wind from either direction.

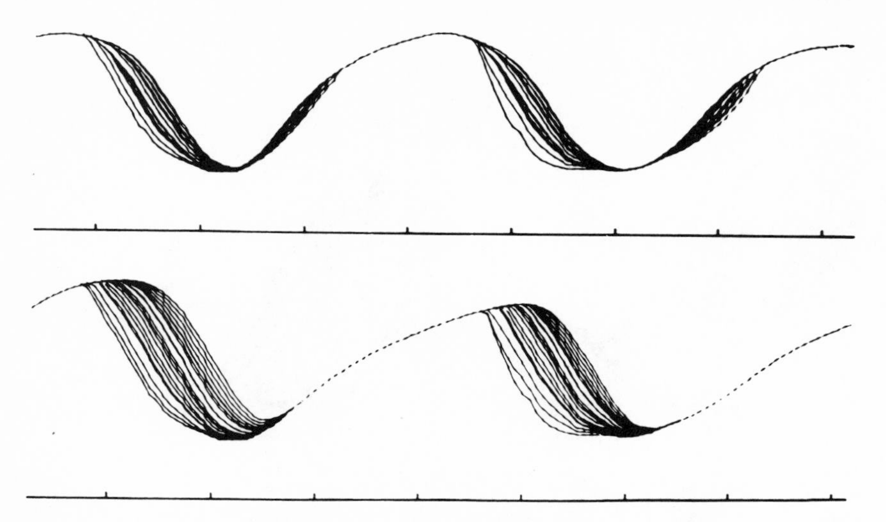

Figure 12. - Wind from either direction with increase in source material after 270 significant intervals.

and to manipulate these variables quickly. With these capabilities, the opportunities for changing dimensions of individual variables and combinations of variables are nearly limitless.

Although this simulation is extremely simplified, it shows in

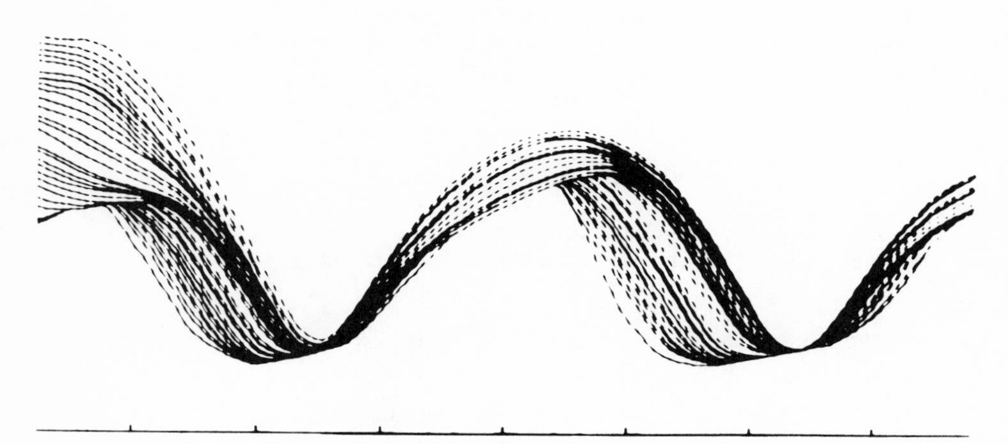

Figure 13. - Reduction in effective wind speed with barrier creating initial lee effect reduced and profiles drawn twice as often.

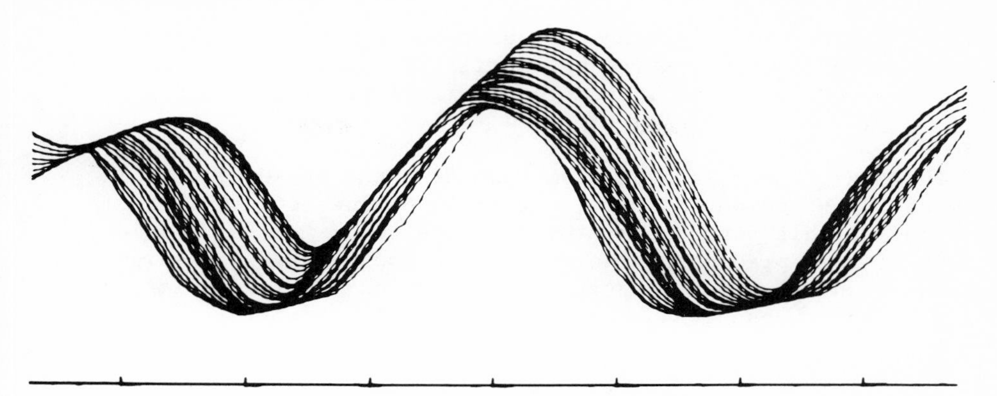

Figure 14. - Wind from two directions with reduction in effective wind speed, reduction in height of barrier creating initial lee effect, and profiles drawn twice as often.

graphical forms some of the factors which are important in shaping sand dunes and suggests the need to consider other environmental factors. It also illustrates some of the processes that have a major influence upon dune migration. It has been shown that wind velocity is of major importance if considering the rapidity of sand migration as are the characteristics of the source supply. The availability of sand for movement is important if considering the change in dune shape, and to a lesser degree, if considering rate of dune migration.

Several variables have been considered and manipulated in the simulation. Such characteristics as wind velocity, slip face and lee effects, and critical wind velocities can be carefully chosen to produce a dynamic dune profile which will approximate reality. It is difficult to determine how well any computer model approaches reality. This simulation is no exception. No tests have yet been conducted with data collected in the field as at this stage it is unlikely that agreement would be high as too many important variables have been omitted.

Further development of the simulation is, of course, required. However, this requires a major move forward in terms of the number of variables involved and the interactions of these variables. Future research requires the introduction of a three-dimensional model, the frictional characteristics of sand grains, variable flow paths around or over dunes, information regarding sediment sizes as opposed to an average size, and moisture and vegetation effects. Also development must include the possibility of dune movement or stabilization under differing conditions of vegetative growth. This in turn will be related to moisture characteristics of the area of dune migration which will influence the movements of various sediment sizes, thus forming a highly interrelated system.

This next stage is a quantum jump from the present, both in terms of labor, computing time, and field evaluation. The model described in this paper indicates that the chance of success in a project of this predicted size is high as the results gained so far, based on only the grossest of information and interrelations, do appear to represent at least a theoretical reality.

REFERENCES

Allen, J. R. L., 1968, Current ripples: their relation to patterns of water and sediment motion: North-Holland Publ. Co., Amsterdam, 433 p.

Bagnold, R. A., 1937, The transport of sand by wind: Geographical Jour., v. 89, p. 409-438.

Bagnold, R. A., 1951, Sand formations in southern Arabia: Geographical Jour., v. 118, p. 78-86.

Bagnold, R. A., 1954, The physics of blown sand and desert dunes (2nd. ed.): Methuen and Co., London, 265 p.

ADDITIONAL REFERENCES

Beheiry, S. A., 1967, Sand forms in the Coachella Valley, southern California: Annals, Assoc. Amer. Geographers, v. 57, no. 1, p. 25-48.

King, C. A. M., and McCullagh, M. J., 1971, A simulation model of a complex recurved spit: Jour. Geology, v. 79, no. 1, p. 22-37.

Norris, R. M., 1966, Barchan dunes of Imperial Valley, California: Jour. Geology, v. 74, no. 3, p. 292-306.

Scheidegger, A. E., 1970, Theoretical geomorphology (2nd. ed.): Springer-Verlag, Berlin, 435 p.

Walton, K., 1969, The arid zones: Aldine, Chicago, Illinois, 175 p.

MATHEMATICAL MODELS FOR SOLUTION RATES OF DIFFERENT-SIZED PARTICLES IN LIQUIDS

Dietrich Marsal

Gewerkschaften Brigitta und Elwerath Erdgas Erdöl

ABSTRACT

This paper introduces a more or less complete quantitative theory of the mechanism of the dissolution of grain aggregates in liquids. The theory is not confined to special grain-size distributions or to special shapes of the grains.

INTRODUCTION

The partial or total dissolution of grains of different size and shape in their own solution or by means of chemical reaction with some type of liquid is a phenomenon that may occur in various geological processes. We remember, for instance, the reduction of grain sizes of flow-agitated or sinking particles in rivers, seas and other waterways; diagenetic processes in the pores of unsolidified sediments; and processes of dissolution with increased temperatures such as occur in rock melts. We further remember analogous processes in laboratory and industrial processing engineering, the separation of minerals by means of dissolution, fractionation in unsaturated solutions in an Atterberg cylinder as well as the dissolution of filter cakes. It also should be mentioned that the mechanism dealt with in this paper also can be applied in the theory of soils.

A general investigation of the dissolution of a grain aggregate, disregarding for now any specific issues, will have to deal in particular with the following problems.

- Which factors influence the change of a grain-size distribution?

- How does the grain-size distribution change with time?

- How does the concentration of the solution change with time?

A typical problem would be: Which particles of a grain aggregate sedimenting in a waterway are dissolved before reaching the bottom?

GRAIN-SIZE REDUCTION DIAGRAM

The fundamental relationships are easy to recognize by means of a so-called grain-size reduction diagram (Fig. 1). The original radii R of the grain aggregate are marked on the abscissa, the associated radii r at any time t are marked on the upward ordinate. Immediately prior to the beginning of dissolution r and R are identical so that the relationship between them is graphically given by the angle bisector of abscissa and upward ordinate. At any later time t the radii of the biggest grains have been reduced from R_{max} to a smaller value r_{max}. All the other grains also have become smaller; for example, the particles with initial radius R_1 are reduced to the radius r_1. All grains of an original radius smaller than R^* have been dissolved.

The lower part of the diagram shows the original grain-number frequency of the aggregate computed from the weight and the original grain radius of each fraction. So the hatched area corresponds to the number of all grains of an original fraction of a radius R_1. The change in grain distribution can be read directly from the figure.

Due to the process of solution the grains of the original radius R_1 have adopted grain size r_1 at the point of time t. Because the number of grains did not change, the grain-number frequency corresponding to r_1 at time t is again indicated by the hatched area.

So we have to determine the isochrones relating R to r at any point of time t. These isochrones depend on the initial grain-size distribution, the shape of the grains and various physical conditions to be considered later. We begin with the phenomena at a single particle.

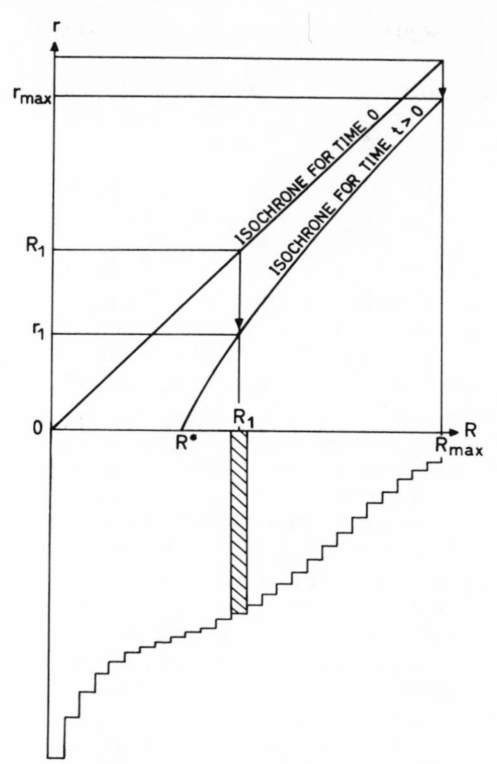

Figure 1. - Scheme of grain-size reduction diagram.

BASIC EQUATIONS

We will first deal with the solution of a particle of radius r [cm] and surface F [cm^2] in its own solution, whereby the grain may have any shape. Let w [cm^3] be the volume of a grain molecule. The removal of dN molecules reduces grain size r by dr and the grain volume by $dV = wdN = Fdr$, so that $dN/dt = -(F/w)(dr/dt)$ molecules migrate per second into the boundary layer between the solid and the surrounding fluid. From this boundary layer the molecules are transported by diffusion. Hence, in the vicinity of the grain surface, Fick's first law may be applied. It states that $J = -D \; grad \; c^*$ molecules leave each square centimeter of the boundary layer per second, D being the coefficient of diffusion [cm^2/s] and c^* the number of dissolved molecules per cm^3 of the boundary layer.

Assuming that at the grain surface the saturation concentration c_s [gram/cm^3] prevails (boundary condition of Nernst), the number of molecules dissolving per second must be equal to the migration rate due to diffusion: $dN/dt = F\bar{J} = -DF\,\overline{grad\ c^*}$, where $\bar{J}$ and $\overline{grad\ c^*}$ signify average values (condition of equilibrium). Hence, it follows that $dr/dt = \omega D\,\overline{grad\ c^*}$. If σ is the mass of a dissolved molecule, ρ its density and c the concentration [gram/cm^3] in the boundary layer, we obtain $\omega = \sigma/\rho$ and $c^* = c/\sigma$, and consequently

$$\frac{dr}{dt} = \frac{D}{\rho}\,\overline{grad\ c} \quad . \tag{1}$$

If the relative velocity between the dissolving particle and the solution is more or less negligible, then the problem of determining $grad\ c$ reduces to a problem of diffusion theory. This situation is dealt with in the last section of this paper.

If convection in the surroundings of the dissolving particle is appreciable, then the region of diffusion will be confined to an enveloping layer of thickness δ with the boundary concentrations c_s and C resp., C being the concentration of the solution [gram/cm^3]. Hence $grad\ c$ may be approximated by $-(c_s-C)/\delta$ for a plane dissolving surface. The gradient related to a grain may be estimated in the following manner. Let F^* be any closed surface within the diffusion layer surrounding the dissolving grain. It is crossed by $F^*\bar{J}$ molecules per second, and this number will depend on time but not much on F^* if δ is small. However, $\bar{J}$ is proportional to $|grad\ c|$ and F^* is proportional to y^2 where y is a mean distance from the midpoint of the grain to F^*. Hence, $|grad\ c|$ is proportional to $1/y^2$ yielding $c-C = (A/y) + B$ with parameters A and B to be determined from the boundary conditions $c=C$ for $y=r$ and $c=C$ for $y = r + \delta$. This yields finally $(\partial c/\partial y)_{y=r} = -(\frac{1}{r} + \frac{1}{\delta})(c_s-C)$.

Now we improve this rough estimation by introducing two empirical shape factors f_1, f_2^*. We obtain the approximation

$$\overline{grad\ c} = -(\frac{f_1}{r} + \frac{f_2^*}{\delta})(c_s-C). \tag{2}$$

The situation $f_1 = f_2^* = 1$ relates to a dissolving sphere, the situation $f_1 = 0$, $f_2^* = 1$ to a dissolving plane; and the choice $f_1 = 0$ leads to an approximation for a grain whose surface is composed of a few planes. Equation (2) will be a good approximation of $\overline{grad\ c}$ if δ is small and it may be poor if δ is large.

As the next step, we have to determine the thickness δ of the diffusion layer. We assume a relative velocity U between the dissolving particle and the solution. At the grain surface, the relative velocity is equal to zero. Within a layer surrounding the particle (boundary layer of Prandtl) the relative velocity assumes all values between 0 and U, being zero at the grain surface. Hence, within Prandtl's layer, all transport phenomena are determined by both diffusion and convection. However, applying Prandtl's theory for a laminar boundary layer and the theory of diffusion to a freely floating dissolving plate, Prandtl's boundary layer may be replaced by a model zone composed of a diffusion layer appropriately chosen and a convection layer superposing the zone of diffusion. The theory yields

$$\delta = 3 \cdot D^{1/3} \cdot \nu^{1/6} \cdot (x/U)^{1/2} , \tag{3}$$

ν being the kinematic viscosity of the liquid and x the distance from the point examined to the point where the flow reaches the plate (Vielstich, 1953; Marsal, 1970). The situation of a turbulent boundary layer is not interesting. It only occurs at extreme high velocities U and if the grain size exceeds the millimeter range.

If the grain is a cube with length L of its edge, then the average value of $\sqrt{x}$ is equal to $(2/3)\sqrt{L}$ yielding $\overline{\delta} = 2 \cdot D^{1/3}\nu^{1/6} \cdot (L/U)^{1/2}$. The surface of a cube is $F = 6L^2$, and its volume change dV equals $3L^2dL$ if L changes from L to $L+dL$. By definition $dV = Fdr$ (cf. first paragraph of this section) yielding $dr = dL/2$. Hence, the application of formulae (1), (2) and (3) with the choice $f_1 = 0$, $f_2{}^* = 1$ results in

$$\frac{dL}{dt} = \rho^{-1} D^{2/3} \nu^{-1/6} (U/L)^{1/2} (c_s - C). \tag{4}$$

In the situation of an irregularly shaped particle the factor 3 of formula (3) has to be changed, and x will be proportional to the radius r if the grain will not change appreciably its shape in the process of solution. So replacing the shape factor $f_2{}^*$ of equation (2) by f_2, we get from (1), (2) and (3),

$$-\frac{1}{c_s - C} \frac{\rho}{D} \frac{dr}{dt} = f_1 r^{-1} + f_2 D^{-1/3} \nu^{-1/6} U^{1/2} r^{-1/2} \tag{5}$$

the shape factors being dependent too on the definition of the grain radius r.

We now will deal with the situation where the grains are dissolved by a reagent of the concentration C_{Re}. Let each arriving molecule immediately react with λ molecules at the surface of the solid substance so that the concentration $c_{Re} = 0$ will always prevail in the phase boundary layer. If the product of the reaction does not influence the kinetics of the process, the relation obtained is analogous to that obtained above, i.e.

$$\frac{dr}{dt} = (\lambda\sigma' D_{Re}/\rho)\ \overline{grad\ c_{Re}}\ ,$$

where σ' is the mass ratio between grain molecule and reagent molecule, D_{Re} is the coefficient of diffusion of the reagent, and ρ is the grain density.

SOLUTION OF DIFFERENT SIZED PARTICLES IN ITS OWN SOLUTION

In the simplest example the concentration C of the solution is independent of location. Then the change in concentration may be calculated directly from the loss of mass of the grain aggregate. (If the process of solution takes place in a reagent, the formulae have to be modified accordingly.)

A grain-size reduction from R to r means that the mass $m(R)dR$ of an original fraction with radii $R...R + dR$ has decreased to $(r/R)^3 m(R)dR$, provided that the density and the shape of the grains have been essentially preserved in the course of dissolution. Hence, if the change of C is exclusively determined by the decrease in volume of the grains, and if we consider a closed system with the volume V and the initial concentration $C(0)\,[C(0) \geq 0]$ of the liquid and initial total mass M of the grain aggregate, then

$$C = C(0) + (M/V) - (1/V) \int_0^{R_{max}} (r/R)^3 m(R)dR. \tag{6}$$

The system (5), (6) is a functional equation with the unknown r. But r depends not only on time t but also on the initial radius R. So the formulae (5), (6) are a nondenumerable system of integro-differential equations. Nevertheless it may be handled numerically for the most important examples.

Case 1: $f_1 = 0$. We put $f_3 = f_2\rho^{-1}D^{2/3}v^{-1/6}U^{1/2}$ and integrate formula (5),

$$R^{3/2} - r^{3/2} = \frac{3}{2}\int_0^t f_3(c_s - C)dt. \tag{7}$$

If the shape factor f_2 does not depend on R, then the right side of equation (7) does not contain R explicitly. In this case the choice of $R = R_{max}$ and $r = r_{max}$ and comparison with (7) yields an equation of the isochrones

$$R^{3/2} - r^{3/2} = R_{max}^{3/2} - r_{max}^{3/2} \,. \tag{8}$$

All grains with initial radii smaller than R^* are dissolved $(r=0)$. So

$$R^{*3/2} = R_{max}^{3/2} - r_{max}^{3/2} \,, \tag{9}$$

and consequently,

$$C = C(0) + (M/V) - (1/V)\int_{R^*}^{R_{max}} (r/R)^3 m(R)dR \tag{10}$$

becomes a function of the largest grain radius r_{max}: $C = C(r_{max})$. Introducing r_{max} into equation (5) yields

$$dr_{max}/dt = -f_3 r_{max}^{-1/2}[c_s - C(r_{max})], \tag{11}$$

and the numerical solution of this differential equation yields r_{max} vs. t solving the problem.

If the shape factor f_2 is unknown, some grain aggregate has to be dissolved completely measuring C, c_s, D, v and U vs. time. Then according to equation (7)

$$R_{max}^{3/2} = \frac{3}{2} f_2 \int_0^{t^*} \rho^{-1} D^{2/3} v^{-1/6} U^{1/2} (c_s - C)dt, \tag{12}$$

t^* being the time interval between the beginning of the dissolution and the disappearance of the last grain. This method avoids the necessity to determine isochrones and the initial grain-size distribution.

Case 2: $f_1 \neq 0$, $D^{-1/3} v^{-1/6} U^{1/2} = const = \phi$. This case is important because usually $D^{-1/3} v^{-1/6}$ is a more or less fixed number; $D^{-1/3}$ is only proportional to the 6th root of absolute temperature, and doubling of v changes $v^{1/6}$ just for 12 percent. Integration of equation (5) yields r vs. r_{max},

$$I(r,R) \equiv \int_r^R \frac{r'dr'}{f_1 + \phi f_2\sqrt{r'}} = \int_{r_{max}}^{R_{max}} \frac{r'dr'}{f_1 + \phi f_2\sqrt{r'}} = I(r_{max}), \tag{13}$$

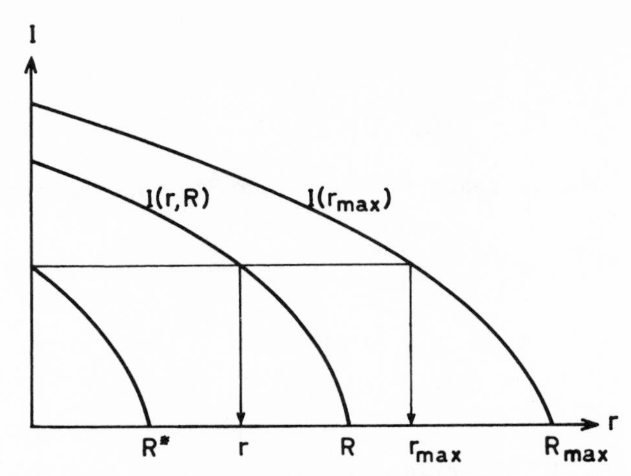

Figure 2. - Determination of r and R^* vs. r_{max} for case 2.

each chosen value of r_{max} giving rise to one isochrone of the grain-reduction diagram. The functions $I(r_{max})$ and $I(r,R)$ are best determined by numerical integration, and $I(r,R) = I(r_{max})$ yields corresponding values of r and r_{max} (Fig. 2). The rest of the analysis runs as shown. The shape factors may be determined by experiment using the equation

$$\int_0^{R_{max}} \frac{r'dr'}{f_1+\phi f_2\sqrt{r'}} = \int_0^{t^*} \frac{D}{\rho}(c_s-C)dt. \tag{14}$$

Case 3: The aggregate is composed of a finite number of classes. Then all grains belonging to the same class have the same radius, and there are as many equations of the type (5) as classes. The integral of formula (6) reduces to a sum. As soon as a class disappears, its differential equation has to be deleted. No problems arise if the parameter ϕ defined as case 2 changes with time.

At last we want to point to the fact that the saturation concentration c_s may change. This will happen if the temperature or the pressure changes or if the chemical composition of the solution changes.

Further, there is no problem of handling the simultaneous solution of several grain aggregates of different chemical composition. In this case each aggregate is related to a system (5),

(6) with concentrations C_1, C_2 etc. and c_{s_1}, c_{s_2} etc., where c_{s_1} depends on C_2, C_3 etc. and c_{s_2} depends on C_1, C_3 etc. and so on.

GRAIN-SIZE REDUCTION IN CASE OF CREEPING FLOW

In a situation of a low velocity U, convection will be more or less negligible. Here we will confine ourselves to spherical symmetry.

If y is the distance from the center of a sphere and r its radius, the diffusion equation $c_t = D[c_{yy} + (2/y)c_y]$ will apply for $t \geq 0$ in the region $r \leq y \leq r+\delta$. The solution of this equation cannot be obtained right away, because the boundary conditions $c = c_s$ for $y = r$ and $c = C$ for $y = r + \delta$ apply to shrinking values of r. If we introduce, however, the nondimensional values $\theta = y/r$ and $\zeta = Dt/r^2$, the diffusion equation will go over into $c_\zeta = c_{\theta\theta} + (2/\theta)c_\theta$ with $c = c_s$ for $\theta = 1$ and $c = C$ for $\theta = 1 + (\delta/r)$ so that one of the boundary conditions has become independent of r. As according to (3) δ/r is proportional to $1/\sqrt{Ur_2}$, δ/r will become so big in the example of a small U and r that we can set $c = C$ for $\theta = \infty$ in a good approximation. If C is fixed at infinity and if we consider the value $\bar{c} = c - C$, we will have to investigate the system

$$\bar{c}_\zeta = \bar{c}_{\theta\theta} + (2/\theta)\bar{c}_\theta$$

$$\theta = 1: \quad \bar{c} = c_s - C \qquad \lim_{\theta \to \infty} \bar{c} = 0 \tag{15}$$

$$\zeta = 0, \; \theta > 1: \quad \bar{c} = 0.$$

Its solution is well known (Carslaw and Jaeger, 1959, p. 247) and supplies

$$grad\ c = \left(\frac{\partial c}{\partial y}\right)_{y=r} = -\left(\frac{1}{r} + \frac{1}{\sqrt{\pi D t}}\right)(c_s - C), \tag{16}$$

so that according to equation (1)

$$\frac{dr}{dt} = -\frac{D}{\rho}\left(\frac{1}{r} + \frac{1}{\sqrt{\pi D t}}\right)(c_s - C). \tag{17}$$

If we integrate the equation (17) and if we insert the identity $r = (r-R) + R$ on the right side of the integrated formula, the result is

$$R^2 - r^2 = (2D/\rho)(c_s - C)\{t + 2R\sqrt{t/\pi D} - \int_0^t \frac{R-r}{\sqrt{\pi D t'}}\, dt'\} \ . \tag{18}$$

At the beginning of the process of dissolution, we have $R >> R-r$ so that the integral in (18) can be neglected.

At a later time the term $\sqrt{\pi Dt}$ becomes larger than r and equation (17) degenerates to $dr/dt = -(D/\rho r)(c_s - C)$ yielding

$$r_0^2 - r^2 = (2D/\rho)(c_s - C)(t - t_0), \tag{19}$$

where

$$r = r_0 \text{ at } t = t_0.$$

A simple example with regard to (19) is supplied by the sedimentation analysis, where an aggregate of small grains is fractionated by making use of their different velocities of sinking. If the velocity of falling is dh/dt and if ρ_F is the density and η [gram/cm·sec] the viscosity of the solution, the relation $dh/dt = 2g(\rho - \rho_F)r^2/9\eta$ will apply according to Stokes with $g = 981$ cm/s^2, provided $r \leq 6 \cdot 10^{-3}$ cm.

Now, if the sinking grain is dissolving during the fall, the equation (17) has to be taken into consideration too. $\sqrt{\pi D}$ is of the order of magnitude of $.01$ cm/s$^{1/2}$ so that $\sqrt{\pi Dt}$ becomes larger than an initial radius $R = 6 \cdot 10^{-3}$ cm as soon as $t > .4$ seconds. Consequently we may use (19) where $r_0 = R$ and $t_0 = 0$ yielding

$$\frac{dh}{dt} = \frac{2g}{9\eta} (\rho - \rho_F) [R^2 - \frac{2D}{\rho} (c_s - C)t] \ , \tag{20}$$

and hence,

$$h = \frac{2g}{9\eta} (\rho - \rho_F)[R^2 t - \frac{D}{\rho}(c_s - C)t^2] \ . \tag{21}$$

Insertion of $t = (R^2 - r^2)/(2D/\rho)(c_s - C)$ yields finally the grain radius r as function of h,

$$(2D/\rho)(c_s - C)h = (g/9\eta)(\rho - \rho_F)(R^4 - r^4) \ . \tag{22}$$

REFERENCES

Carslaw, H. S., and Jaeger, J. C., 1959, Conduction of heat in solids: Oxford Univ. Press, 510 p.

Marsal, D., 1970, Die Auflosung von Korngemengen: Habilitationsschrift, Universitat Stuttgart, 50 p.

Vielstich, W., 1953, Der Zusammenhang zwischen Nernstscher Diffusionsschicht und Prandtlscher Stromungsgrenzschicht: Zeitschrift fur Elektrochemie, v. 57, no. 8, p. 646-655.

APPENDIX

List of frequently used symbols

C	[g/cm^3]	concentration of the solution
c_s	[g/cm^3]	saturation concentration
D	[cm^2/s]	coefficient of diffusion
f_1, f_2		dimensionless shape factors
r	[cm]	grain radius
R	[cm]	initial radius
t	[s]	time
U	[cm/s]	flow velocity
δ	[cm]	thickness of diffusion layer
ν	[cm^2/s]	kinematic viscosity
ρ	[g/cm^3]	density of the grains

A SIMPLE QUANTITATIVE TECHNIQUE FOR COMPARING CYCLICALLY DEPOSITED SUCCESSIONS

William A. Read and Daniel F. Merriam

Kansas Geological Survey and Syracuse University

ABSTRACT

This paper introduces a technique for comparing cyclically deposited sedimentary sequences. Each section is coded into a limited number of lithological states and a first-order transition probability matrix is constructed. The elements of this matrix then are rearranged to form a vector. Sections are compared by calculating product-moment correlation coefficients for the appropriate vectors. Single- and unweighted average-linkage cluster analysis are used to cluster the sections on the basis of these correlation coefficients.

The method has been used to classify two different sets of succession data and in each situation the results are readily interpretable in terms of depositional environments. One data set is derived from cored borehole sections through the same uniform paralic-facies succession in the upper part of the Limestone Coal Group (Namurian, E_1) and located in the small Kincardine Basin, east of Stirling in central Scotland. The other data set, which is coded in less detail, is derived from a varied series of sections in the Carboniferous of Britain and the USA.

INTRODUCTION

The objective of this paper is to describe the preliminary results of using a comparatively simple quantitative technique to compare cyclically deposited successions. These results seem, in general, to make better sense from a geological and sedimentological standpoint than those obtained by applying cross-association to the same data and indicate that the technique has potential for further development. The mathematical model used to represent the successions is that of a first-order Markov chain. This mathematical model has been used because it is relatively simple and readily comprehensible to nonmathematical geologists. It is fully realized also that a second-, or higher, order Markov chain may well provide a better model for many of the successions studied (Schwarzacher, 1967, 1969; Doveton, 1971). The conceptual depositional model which seems to be most readily applicable to the majority of the sections studied is that of repeated deltaic cycles (Coleman and Gagliano, 1964; Read, 1969b), in which the clastic wedges formed by delta lobes are built, colonized by vegetation and ultimately abandoned. In this paper an attempt is made to link the mathematical and conceptual models and to assess the sedimentological results in some detail.

Only a decade ago cyclically deposited successions were generally compared by considering the idealized fully developed cyclothem which had been constructed intuitively for each succession (Robertson, 1948; Weller, 1958). The considerable disadvantages inherent in the approach have been described by Duff, Hallam, and Walton (1967, p. 3-8).

Two mainly independent lines of investigation have replaced the fully developed cyclothem concept. The first, which tends to be followed by mathematical geologists, analyzes the sequences of lithologies which are actually observed. This approach, which owes much to the pioneer work of Duff and Walton (1962), has recently received considerable stimulus from the application of the Markov-process model (Krumbein, 1967; Krumbein and Scherer, 1970; Schwarzacher, 1967, 1969).

The other line of investigation, which tends to be followed by classical sedimentologists, is concerned with the recognition of sequences of despoitional environments. It originated from detailed sedimentological investigations of modern sedimentary environments, particularly deltaic environments (Fisk and others, 1954; Allen, 1964, 1965b, 1970; Coleman and Gagliano, 1964, 1965; Frazier, 1967; Oomkens, 1967, 1970; Kanes, 1970), and the recognition of these environments in ancient cyclically deposited rocks, especially in Britain (Moore, 1958, 1959; Allen, 1965a; Elliott, 1968, 1969; Read, 1965, 1969a; Read, Dean, and Cole, 1971) and in the USA (Williams and Ferm, 1964; Ferm and Cavaroc, 1968, 1969; Ferm, 1970; Fisher and McGowen, 1967, 1969; Wanless and others, 1970).

A few workers (Potter and Blakely, 1968; Read, 1969b; Selley, 1970) have made attempts to combine these two lines of approach and it seems likely that in the future increasing attention will be devoted to expressing, in mathematical terms, the vertical and horizontal relationships between sedimentary environments, as opposed to pure lithologies.

Particular cyclical successions have been compared and classified qualitatively in terms of the sedimentary environments which are represented (Fisher and others, 1969) but comparatively few attempts have been made to compare particular cyclical successions quantitatively. However, some workers (Sackin, Sneath, and Merriam, 1965; Merriam and Sneath, 1967; Merriam, 1970; Read, 1970; Read and Sackin, in press) have used cross-association techniques to compare a series of cyclical successions which have been coded into a limited number of lithological states. The most commonly used technique has been the calculation of cross-association similarity coefficients (Sackin, Sneath, and Merriam, 1965, p. 13-16; Harbaugh and Merriam, 1968, p. 171-173) for each pair of successions and then use cluster analysis (Sokal and Sneath, 1963, p. 179-182; Parks, 1966; Harbaugh and Merriam, 1968, p. 174-179) to emphasize the relationships between sections.

This paper describes an alternative approach in which sections are compared by means of the first-order transition probabilities of passage upward from one specified lithology to another. As in the cross-association studies, the successions are first coded into a limited number of lithological states. A detailed investigation of the relative advantages and disadvantages of using this method as compared to the calculation of cross-association similarity indices falls outside the scope of this paper but is being studied separately.

SOURCES OF DATA AND METHOD OF CODING

The succession data used in the paper are simple successions of lithological members coded into a limited number of states. No account has been taken of the thickness of each member and no multistorey lithologies are recognized (Read, 1969b, p. 202). Thus, it is not considered possible for a given lithological state to pass upwards into the same lithological state.

The technique used in the present study was tested on two different sets of succession data: the first set, here termed the Kincardine Basin data set, comprises a series of sections through the same uniform succession in a small area, whereas the second set, here termed the UK-US data set, comprises a heterogeneous series of Carboniferous sections collected from a wide range of facies types and stratigraphical intervals in various parts of Britain and the USA. The succession data which were designated the List 2 data in the cross-association study by Read and Sackin (in press) constitute the Kincardine Basin data set. These data were extracted from the logs of fifteen cored boreholes through the upper part of the Limestone Coal Group of central Scotland, which is a coal-bearing, paralic-facies sequence of marked overall uniformity (Read, 1969a, p. 332; 1969b, p. 200). All the bores lie within a single basin of deposition (the Kincardine Basin) which is only about 15 miles (24 km) wide and lies immediately east of Stirling (Read and Dean, 1967, fig. 1). The name and location of these boreholes are listed in Table 1 and are designated by the terms A1 to A15. These terms are used instead of the borehole names throughout the text because of the lengths of the names and also in order to facilitate comparison with other publications in which the same terms are used (Read and Dean, 1967, 1968; Read and Sackin, in press). The cores of these boreholes were examined in considerable detail and in a uniform manner by geologists from the Edinburgh Office of the Institute of Geological Sciences. The succession data were extracted from the borehole logs in the public files of that office.

The lithological data were coded into five states, namely mudstone, siltstone, sandstone, seatrock and coal. The coded successions have been listed by Read (1970, table 34). All five states are well represented in each of the fifteen sections. The thinnest section contained more than 100 lithological members and was more than 436 feet (133 m) thick.

Table 1. - Names and geographical and stratigraphical locations of sections.

I. KINCARDINE BASIN DATA SET
(All in upper part of Limestone Coal Group, Namurian, E_1. British National Grid References are given in brackets)

Section No.	Name
A1	Powis Mains No. 1 Bore, 1959 (NS 822958)
A2	Torwood No. 1 Bore, 1927-8 (NS 835843)
A3	Torwood Bore, 1960-1 (NS 838849)
A4	Glenbervie No. 4 Bore, 1926 (NS 850857)
A5	Tullibody No. 2 Bore, 1934-6 (NS 869938)
A6	Doll Mill Bore, 1955 (NS 875881)
A7	Mossneuk Bore, 1950-2 (NS 872861)
A8	South Letham No. 1 Bore, 1952 (NS 886853)
A9	Kincardine Bridge Bore, 1951-2, (NS 917872)
A10	Orchardhead Bore, 1956 (NS 924841)
A11	Grangemouth Dock Bore, 1956-7 (NS 951839)
A12	Righead Bore, 1953-6 (NS 972882)

Section No.	Name
A13	Culross No. 2 Bore, 1957 (NS 983859)
A14	Solsgirth Bore, 1941-61 (NS 997948)
A15	Shepherdlands Bore, 1933-4 (NT 006902)

II. UK-US DATA SET

(National Grid References for UK sections and township and range notation for US sections are given in brackets.)

Name	Geographic and stratigraphic location
Askrigg	UK, Yorkshire. Stream section NE of Askrigg (SD 952922). Middle Limestone Group: Visean, P_1.
Dosthill	UK, Warwickshire. Baggeridge Brick Co. open pit N of Kingsbury (SP 218995). Middle Coal Measures: Westphalian B and C.
Dunbar	UK, East Lothian. Associated Portland Cement Co. quarry (NT 708761). Upper Oil Shale Group and Lower Limestone Group: Visean, P_1 and P_2.
Foggermount	UK, Stirlingshire. Caledon Coal Co. open cast working (NS 923732). Lower Coal Measures: Westphalian A.
Ingleton	UK, Yorkshire. Section in Aspland Beck (NS 660719). Lower Coal Measures: Westphalian A.
Joppa	UK, Midlothian. Shore section (NT 322734). Passage Group and Lower Coal Measures: Westphalian A.
Neepsend	UK, Yorkshire. Sheffield Brick Co. open pit (SK 351890). Lower Coal Measures: Westphalian A.
Overseal	UK, Leicestershire. Robinson and Dowlers Pipe Clay pits (SK 298162). Middle Coal Measures: Westphalian B and C.
Seaton Sluice	UK, Northumberland. Shore section in Hartley Bay (NZ 345756). Lower and Middle Coal Measures: Westphalian A and B.
Upper Bannock	UK, Stirlingshire. Stream section in Bannock Burn (NS 746877). Topmost Calciferous Sandstone Measures and Lower Limestone Group: Visean, P_1 and P_2.

Name	Geographic and stratigraphic location
Allegheny River	US, Pennsylvania, Armstrong County. Railroad cut on W side of Allegheny at Kittaning. Pottsville and Allegheny Groups: Atokan and Desmoinesian.
Athens	US, Ohio, Athens County. Roadcuts near Nelsonville (NW 23 and SW 24, T. 12 N., R. 15 W.). Allegheny and Conemaugh Groups: Desmoinesian and Missourian.
Blue Mound	US, Kansas Douglas County. Roadcuts (SE c sec. 21, T. 13 S., T. 20 E.). Douglas Group: Virgilian.
Bonner Springs	US, Kansas, Wyandotte County. Roadcut in Kansas Turnpike, mile 13.4 (W2 sec. 18, T. 11 S., R. 23 E.). Lansing, Kansas City and Douglas Groups: Missourian and Virgilian.
Chelyan	US, West Virginia, Kanawha County. Roadcut in West Virginia Turnpike, mile 74. Kanawha Group: Atokan.
Columbia	US, Missouri, Boone County. Columbia Brick and Tile Co. open pits (NE SW sec. 8, T. 48 N., R. 12 W.). Cherokee Group: Desmoinesian.
Francis	US, Oklahoma, Pontoc County. Roadcuts (sec. 3, and 4, T. 4 N., R. 7 E.). Wewoka, Holdenville and Seminole Formations: Desmoinesian and Missourian.
Miller	US, Kansas, Lyon County. Roadcut in Kansas Turnpike, mile 88.7 (NE SE sec. 13, T. 16 S., R. 12 E.). Wabaunsee Group: Virgilian.
Mulberry	US, Kansas, Crawford County. Strip pits (NW sec. 24, T. 28 S., R. 25 E.). Cherokee Group: Desmoinesian.
Valencia	US, Kansas, Shawnee County. Roadcuts near Topeka (SW SW sec. 20 and NW NW sec. 29, T. 11 S., R. 14 E.). Wabaunsee Group: Virgilian.
A1 and A8	See details under Kincardine Basin data set.

The UK-US data set comprises somewhat simplified versions of the 20 sets of coded succession data investigated by cross-association by Merriam (1970, figs. 1, 2 and 4, tables 1 and 2), plus recoded versions of two of the sections in the Kincardine Basin data set, making a total of 22 sections. The original 20 sets of succession data, 10 from Britain and 10 from the USA, were all derived from surface exposures which had been logged, or relogged, by Merriam, thus ensuring a reasonably uniform method of extracting the basic succession data. The names, geographical locations and stratigraphical locations of these sections are listed in Table 1.

In order to be fully effective, the technique used in this paper requires a reasonably high proportion of nonzero values to be present in the transition probability matrix. Consequently the original seven lithological states recognized by Merriam (1970, p. 248, fig. 5) had to be reduced to four, as follows: (1) limestone (unchanged), (2) mudstone (nonfossiliferous shale plus fossiliferous shale), (3) sandy beds (siltstone plus sandstone), and (4) vegetation beds (seatearth or underclay plus coal). This had the somewhat unfortunate effect of reducing the amount of lithological detail recorded but it did eliminate the difficulties found to be inherent in trying to distinguish nonfossiliferous from fossiliferous shales (Merriam, 1970, p. 249).

The two additional sections which were added to the UK-US data set were those found to be most dissimilar in the Kincardine Basin data set. These were recoded in the same simplified form as the other 20 sections and were added to allow some comparison to be made between the two data sets and also to investigate the importance of the level of detail used in coding sections. The recognition of a large number of states, however desirable from a sedimentological standpoint, commonly tends to obscure simple relationships and may over-emphasize minor differences, whereas if too few states are used in coding, detailed relationships are lost and basically different sequences may be grouped together (Read, 1969b, p. 202, 214; Merriam, 1970, p. 248-249). In this study the authors have deliberately decided to run the risk of recognizing too few rather than too many states.

METHOD

A first-order transition probability matrix was constructed for each coded section. Two different approaches are possible. The first is to include only those upward transitions thought to be of particular significance and the second is to include all possible transitions between the states recognized. In general the second approach was followed but, in the analysis of the Kincardine Basin data set, any transition which had a probability of less than an arbitrary level of 0.10 in all fifteen sections was eliminated. Three transitions were accordingly omitted, namely those from mudstone to coal, siltstone to coal and sandstone to coal. Because no lithological state was allowed to pass upwards into itself, zeros appeared in the principal diagonals of all the transition probability matrices and these diagonal elements were eliminated in both data sets. The matrices were then converted to vectors so that a comparison could be made between corresponding transition probabilities in different sections. Thus the vectors derived from the Kincardine Basin data set contained 17 elements and those from the UK-US data set contained 12 elements.

The vectors of transition probabilities for each pair of sections then were used to calculate a product-moment correlation coefficient for that pair of sections and two lower half matrices of correlation coefficients were compiled. Excluding the elements (with unit values) in the principal diagonal the half correlation matrix for the Kincardine Basin data set contained 105 elements and that for the UK-US data set contained 231 elements. A rather undesirable feature of the transition probability data is that all the elements in a single row of the original transition probability matrix for each section are constrained to sum to unity, thus forming a series of closed systems. This feature tends to lower the correlation coefficient values and can give rise to negative values (Chayes, 1960; Krumbein, 1962) but in this study negative values are absent in the correlation matrix for the Kincardine Basin data and are comparatively few and low in absolute value in the correlation matrix for the UK-US data. Thus these negative values had relatively little effect on the final results.

Cluster analyses were used to emphasize relationships between sections, using the two half matrices of correlation coefficients (exclusive of the elements in the principal diagonals) as input. Both single-linkage and unweighted-average linkage methods then were applied to each set of correlation coefficient data (Sokal and Sneath, 1963, p. 180-182) using the TAXON program from the NTSYS series of computer programs (Merriam, 1970, p. 253, footnote). The MXCOMP program from the same series then was employed to calculate cophenetic correlation coefficients, which were used to assess the effectiveness of the cluster analysis dendrograms in representing the correlation coefficient data (Sokal and Sneath, 1963, p. 189).

RESULTS

Kincardine Basin Data Set

Table 2 shows the transition probabilities for the five-state

Table 2. - Transition probabilities for Kincardine Basin data set.

Transitions	Sections (For names see Table 1)														
	A1	A2	A3	A4	A5	A6	A7	A8	A9	A10	A11	A12	A13	A14	A15
Mudstone to siltstone	0.13	0.47	0.29	0.22	0.43	0.58	0.40	0.73	0.59	0.75	0.80	0.47	0.63	0.37	0.67
Mudstone to sandstone	0.83	0.42	0.54	0.67	0.43	0.29	0.40	0.14	0.31	0.13	0.12	0.47	0.21	0.41	0.21
Mudstone to seatrock	0.04	0.11	0.17	0.11	0.11	0.13	0.20	0.09	0.10	0.08	0.08	0.15	0.13	0.22	0.13
Siltstone to mudstone	0.50	0.13	0.08	0.15	0.14	0.13	0.17	0.10	0.16	0.03	0.05	0.21	0.08	0.26	0.06
Siltstone to sandstone	0.50	0.44	0.83	0.45	0.48	0.61	0.54	0.59	0.68	0.67	0.65	0.54	0.50	0.43	0.52
Siltstone to seatrock	0.00	0.38	0.08	0.40	0.38	0.26	0.29	0.31	0.16	0.27	0.30	0.25	0.42	0.30	0.35
Sandstone to mudstone	0.11	0.00	0.04	0.00	0.03	0.06	0.04	0.09	0.05	0.07	0.02	0.06	0.00	0.09	0.03
Sandstone to siltstone	0.04	0.05	0.16	0.31	0.21	0.19	0.22	0.41	0.26	0.22	0.27	0.24	0.33	0.15	0.28
Sandstone to seatrock	0.86	0.95	0.80	0.69	0.73	0.75	0.74	0.50	0.68	0.70	0.71	0.71	0.67	0.73	0.66
Seatrock to mudstone	0.25	0.15	0.08	0.11	0.20	0.11	0.24	0.24	0.20	0.06	0.07	0.05	0.06	0.14	0.19
Seatrock to siltstone	0.00	0.02	0.00	0.14	0.15	0.07	0.18	0.12	0.07	0.00	0.05	0.03	0.08	0.05	0.02
Seatrock to sandstone	0.13	0.13	0.03	0.14	0.10	0.16	0.03	0.24	0.10	0.06	0.20	0.08	0.12	0.18	0.19
Seatrock to coal	0.63	0.70	0.89	0.60	0.56	0.67	0.56	0.41	0.63	0.88	0.68	0.85	0.73	0.64	0.60
Coal to mudstone	0.55	0.33	0.61	0.52	0.60	0.45	0.68	0.53	0.50	0.67	0.50	0.61	0.51	0.45	0.48
Coal to siltstone	0.00	0.15	0.03	0.10	0.20	0.24	0.11	0.27	0.08	0.20	0.08	0.12	0.19	0.17	0.17
Coal to sandstone	0.10	0.03	0.03	0.14	0.08	0.07	0.11	0.00	0.15	0.03	0.08	0.06	0.08	0.14	0.00
Coal to seatrock	0.35	0.48	0.33	0.24	0.12	0.24	0.11	0.20	0.27	0.10	0.34	0.21	0.22	0.24	0.34

coded data derived from the Kincardine Basin sections. Each column represents the vector of transition probabilities for a particular section. Inspection of the rows of this matrix of transition probabilities reveals that, in general, the transition probabilities for most pairs of lithological states do not differ greatly from section to section. Notable exceptions however are the transition probabilities from mudstone to siltstone and from mudstone to sandstone.

Table 3 shows the 15 x 15 lower half matrix of correlation coefficients derived from each pair of column vectors in Table 2. Thus each element in Table 3 gives an assessment of the overall similarity between the transition probabilities for particular pairs of sections. Seventy-five of a total of 105 values are 0.80 or more and 31 are 0.90 or more, indicating a fairly high level of similarity between most of the sections. The least similar sections are A1 (Powis Mains No. 1 Bore) and A8 (South Letham No. 1 Bore) which have a correlation coefficient value of only 0.27. The succession data from these two sections were recoded and then included in the UK-US data set.

Table 3. - Correlation matrix, derived from Table 2, for Kincardine Basin sections.

	A1	A2	A3	A4	A5	A6	A7	A8	A9	A10	A11	A12	A13	A14	A15
A1	1.00														
A2	0.73	1.00													
A3	0.82	0.83	1.00												
A4	0.80	0.81	0.86	1.00											
A5	0.70	0.84	0.85	0.89	1.00										
A6	0.63	0.90	0.89	0.76	0.91	1.00									
A7	0.72	0.78	0.88	0.86	0.97	0.87	1.00								
A8	0.27	0.60	0.60	0.49	0.75	0.83	0.71	1.00							
A9	0.68	0.83	0.91	0.75	0.88	0.96	0.89	0.82	1.00						
A10	0.49	0.76	0.82	0.67	0.87	0.95	0.86	0.87	0.93	1.00					
A11	0.45	0.81	0.78	0.63	0.81	0.94	0.78	0.90	0.94	0.94	1.00				
A12	0.76	0.85	0.94	0.88	0.92	0.93	0.91	0.69	0.92	0.91	0.84	1.00			
A13	0.47	0.83	0.80	0.75	0.89	0.95	0.85	0.86	0.90	0.96	0.95	0.91	1.00		
A14	0.81	0.93	0.91	0.87	0.92	0.92	0.89	0.61	0.87	0.84	0.80	0.96	0.86	1.00	
A15	0.49	0.86	0.79	0.69	0.86	0.95	0.81	0.91	0.92	0.93	0.97	0.86	0.96	0.83	1.00

Figure 1 shows the dendrogram for the single-linkage cluster analysis of the data in Table 3. When the cophenetic values shown by this dendrogram were compared with the values in Table 3, a cophenetic correlation coefficient of 0.71 was obtained. This value is not particularly high (Merriam, 1970, p. 253), suggesting that the dendrogram may not necessarily provide a completely adequate representation of the correlation coefficient data.

Single-linkage cluster analysis tends to form loose, straggling clusters which join together readily (Sokal and Sneath, 1963, p. 190). This tendency is well illustrated in Figure 1 in which all fifteen sections are grouped together at the comparatively high value of 0.81. Between this value and 0.89 there are two distinct

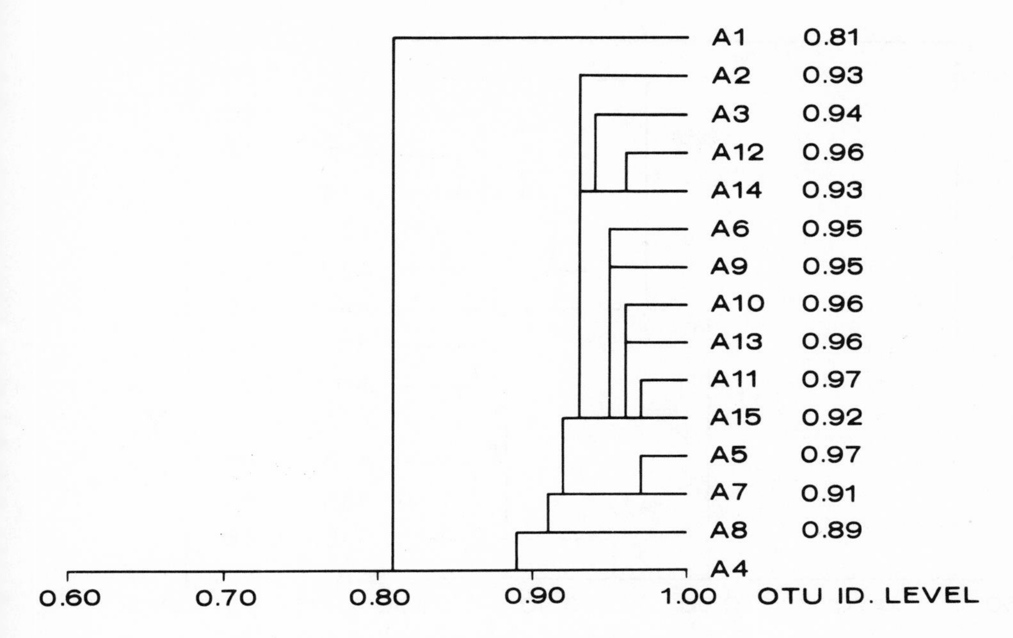

Figure 1. - Dendrogram showing single-linkage cluster analysis of correlation coefficients in Table 3. Kincardine Basin data set.

clusters, one of which comprises Section A1 alone whereas the other comprises all the remaining 14 sections. Reference to the data in Table 3 shows that the transition probability for mudstone up to siltstone is much lower in A1 than in the other sections and the transition probabilities from mudstone to sandstone and siltstone to mudstone are much higher.

Above a value of 0.93 the second, larger cluster splits into several smaller clusters which are somewhat difficult to interpret geologically and so have not been described in detail.

Figure 2 shows the dendrogram for the unweighted average-linkage cluster analysis of the data in Table 3. The cophenetic correlation coefficient value remains at 0.71 so that this dendrogram too may not provide a completely adequate representation of the correlation coefficient data. However, unweighted average-linkage cluster analysis tends to form compact clusters which merge less easily than those of single-linkage cluster analysis. The more gradual clustering shown in Figure 2 can be more readily interpreted than the clustering shown in Figure 1.

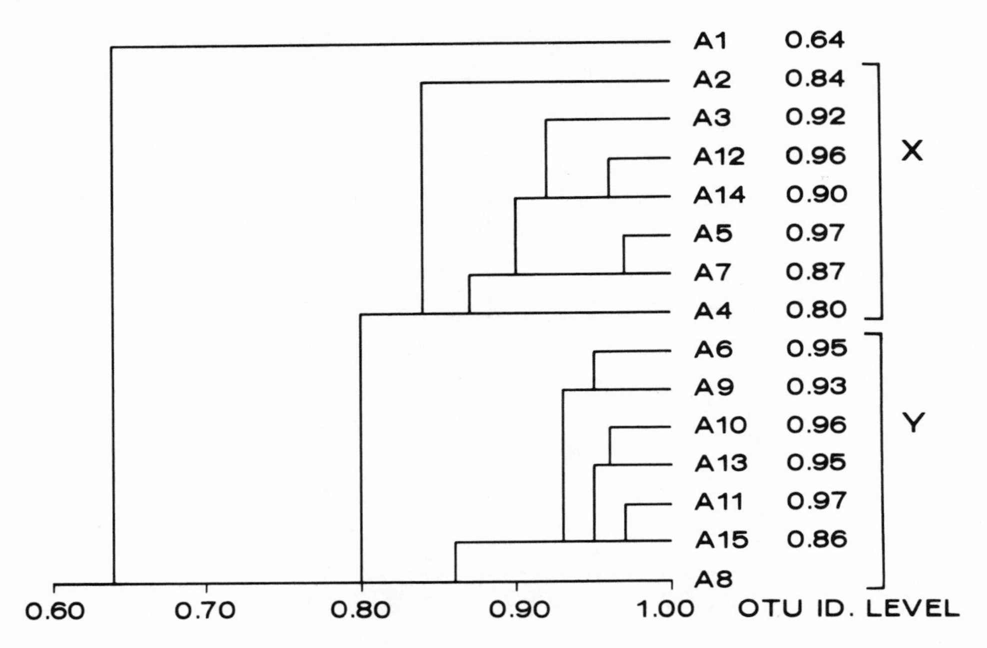

Figure 2. - Dendrogram showing unweighted average-linkage cluster analysis of correlation coefficients in Table 3. Kincardine Basin data set.

The value at which all fifteen sections are grouped together is 0.64. Between this value and 0.80 the sections are grouped into the two major clusters already noted in Figure 1, namely Section A1 above and then all the remaining sections. Above a value of 0.80 the larger cluster splits into two subsidiary clusters of roughly equal size. The first of these, which has been marked 'X' on Figure 2, comprises sections A2-5, A7, A12 and A14. In all of these, apart from Section A2, the probability values for the upward transition from mudstone to siltstone are less than or equal to those for the mudstone to sandstone transition. In Section A2, which splits from the other sections in the 'X' cluster at a level of 0.84, the former probability is slightly higher than the latter and this section also has unusually high probabilities for the transitions from sandstone to seatrock and from coal to seatrock.

The second subsidiary cluster, marked 'Y' on Figure 2, comprises Sections A6, A8-11, A13 and A15. In all of these the transition probability values for mudstone to siltstone exceed those for mudstone to sandstone. In Section A8, which splits from the other members of the 'Y' cluster at a level of 0.86, there is a lower

Table 4. - Transition probabilities for UK-US data set.

Transitions	SECTIONS																					
	Askrigg	Dosthill	Dunbar	Foggermount	Ingleton	Joppa	Neepsend	Overseal	Seaton Sluice	Upper Bannock	Allegheny River	Athens	Blue Mound	Bonner Springs	Chelyan	Columbia	Francis	Miller	Mulberry	Valencia	A1	A8
Limestone to mudstone	1.00	0.00	0.60	0.00	0.00	0.00	0.00	0.00	0.00	0.83	1.00	1.00	0.67	1.00	1.00	0.50	1.00	1.00	0.33	1.00	0.00	0.00
Limestone to sandy beds	0.00	0.00	0.20	0.00	0.00	0.00	0.00	0.00	0.00	0.17	0.00	0.00	0.33	0.00	0.00	0.00	0.00	0.00	0.00	0.00	0.00	0.00
Limestone to vegetation beds	0.00	0.00	0.20	0.00	0.00	0.00	0.00	0.00	0.00	0.00	0.00	0.00	0.00	0.00	0.00	0.50	0.00	0.00	0.67	0.00	0.00	0.00
Mudstone to limestone	0.00	0.00	0.80	0.00	0.00	0.00	0.00	0.00	0.00	0.56	0.10	0.14	0.50	0.56	0.33	0.78	0.29	0.89	0.75	0.88	0.00	0.00
Mudstone to sandy beds	1.00	0.57	0.20	1.00	0.60	0.67	1.00	0.33	0.80	0.33	0.40	0.14	0.00	0.44	0.17	0.00	0.71	0.11	0.13	0.13	0.96	0.86
Mudstone to vegetation beds	0.00	0.43	0.00	0.00	0.40	0.33	0.00	0.67	0.20	0.11	0.50	0.71	0.50	0.00	0.50	0.22	0.00	0.00	0.13	0.00	0.04	0.14
Sandy beds to limestone	0.25	0.00	0.33	0.00	0.00	0.00	0.00	0.00	0.00	0.20	0.00	0.00	0.00	0.33	0.00	0.00	0.00	0.00	0.00	0.00	0.00	0.00
Sandy beds to mudstone	0.75	0.75	0.33	0.40	0.67	0.60	0.80	0.00	0.80	0.60	0.60	1.00	0.33	0.67	1.00	0.00	1.00	1.00	1.00	1.00	0.17	0.19
Sandy beds to vegetation beds	0.00	0.25	0.33	0.60	0.33	0.40	0.20	1.00	0.20	0.20	0.40	0.00	0.67	0.00	0.00	0.00	0.00	0.00	0.00	0.00	0.83	0.81
Vegetation beds to limestone	0.00	0.00	0.00	0.00	0.00	0.00	0.00	0.00	0.00	0.00	0.00	0.00	0.00	0.00	0.00	0.00	0.00	0.00	0.17	0.00	0.00	0.00
Vegetation beds to mudstone	0.00	1.00	1.00	1.00	1.00	0.75	0.50	1.00	1.00	0.50	1.00	0.60	0.50	0.00	0.75	1.00	0.00	0.00	0.83	0.00	0.76	0.50
Vegetation beds to sandy beds	0.00	0.00	0.00	0.00	0.00	0.25	0.50	0.00	0.00	0.50	0.00	0.40	0.50	0.00	0.25	0.00	0.00	0.00	0.00	0.00	0.24	0.50

probability of seatrock passing up into coal than in the other sections.

UK-US Data Set

Table 4 shows the matrix of transition probabilities for the four-state coded data in the UK-US data set. In sharp contrast to Table 2, almost every row contains markedly different transition probability values, stressing the more heterogeneous nature of the basic succession data. A fairly high proportion of the sections contain only three of the four lithological states recognized so that, unlike Table 2, there are a fairly large number of zero values in the matrix. A further difficulty is that the total numbers of lithological members recognized in some of the shorter sections are considerably less than the total number of members in the shortest of the Kincardine Basin Sections. Thus some of the transition probabilities are based on a fairly small number of transitions. It is interesting to note how closely the transition probabilities for the recoded data from Sections A1 and A8 correspond, in contrast to the differences between the column vectors

Table 5. - Correlation matrix, derived from Table 4, for UK-US sections

	Askrigg	Dosthill	Dunbar	Foggermount	Ingleton	Joppa	Neepsend	Overseal	Seaton Sluice	Upper Bannock	Allegheny River
Askrigg	1.00										
Dosthill	0.24	1.00									
Dunbar	0.09	0.36	1.00								
Foggermount	0.30	0.81	0.38	1.00							
Ingleton	0.22	1.00	0.37	0.86	1.00						
Joppa	0.28	0.94	0.21	0.89	0.95	1.00					
Neepsend	0.52	0.69	0.03	0.75	0.69	0.84	1.00				
Overseal	-0.23	0.62	0.28	0.66	0.68	0.64	0.17	1.00			
Seaton Sluice	0.37	0.97	0.37	0.88	0.97	0.94	0.81	0.50	1.00		
Upper Bannock	0.55	0.23	0.59	0.16	0.21	0.23	0.34	-0.10	0.28	1.00	
Allegheny River	0.49	0.67	0.57	0.52	0.67	0.57	0.30	0.51	0.61	0.62	1.00
Athens	0.50	0.48	0.25	0.08	0.42	0.37	0.28	0.07	0.39	0.70	0.78
Blue Mound	-0.07	0.13	0.39	0.03	0.14	0.15	-0.10	0.43	0.00	0.58	0.54
Bonner Springs	0.81	0.00	0.39	-0.03	-0.03	-0.05	0.17	-0.39	0.10	0.76	0.44
Chelyan	0.52	0.53	0.47	0.19	0.48	0.40	0.30	0.08	0.48	0.79	0.83
Columbia	-0.15	0.25	0.82	0.16	0.25	0.06	-0.18	0.25	0.20	0.36	0.46
Francis	0.92	0.25	0.21	0.20	0.22	0.25	0.47	-0.28	0.36	0.70	0.53
Miller	0.57	0.04	0.44	-0.13	-0.01	-0.06	0.09	-0.37	0.08	0.77	0.40
Mulberry	0.12	0.49	0.60	0.22	0.44	0.29	0.23	-0.03	0.48	0.40	0.39
Valencia	0.58	0.04	0.44	-0.12	0.00	-0.05	0.10	-0.37	0.08	0.77	0.41
A1	0.20	0.63	0.22	0.94	0.70	0.81	0.69	0.71	0.69	0.08	0.38
A8	0.17	0.50	0.00	0.80	0.57	0.76	0.71	0.64	0.56	0.08	0.25

for these sections in Table 2. This, however, is partly due to the fact that all the transitions involving limestone have zero values in both sections.

Table 5 shows the 22 x 22 lower half matrix of correlation coefficients derived from each pair of columns in Table 4. The values are more variable and are generally considerably lower than those for the Kincardine Basin sections shown in Table 3. Some negative values also appear but the absolute values of these are low. The correlation coefficient for the recoded versions of the A1 and A8 successions is 0.95, thus demonstrating the marked effect of the detail in which coding is accomplished.

Figure 3 shows the dendrogram produced by single-linkage cluster analysis of the correlation coefficient data in Table 5. The cophenetic correlation coefficient for this dendrogram is only 0.68 which is again rather low.

All 22 sections are grouped together at a level of 0.58. At

Blue Mound	Bonner Springs	Chelyan	Columbia	Francis	Miller	Mulberry	Valencia	A1	A8
1.00									
0.17	1.00								
0.51	0.63	1.00							
0.32	0.14	0.40	1.00						
0.12	0.90	0.69	-0.02	1.00					
0.35	0.89	0.71	0.26	0.82	1.00				
0.06	0.33	0.59	0.64	0.38	0.57	1.00			
0.35	0.90	0.71	0.26	0.82	1.00	0.57	1.00		
0.12	-0.15	-0.01	0.01	0.06	-0.27	-0.05	-0.26	1.00	
0.18	-0.19	-0.07	-0.17	0.03	-0.30	-0.20	-0.29	0.95	1.00

a level of 0.70 there are two definite clusters of roughly equal size, marked on Figure 3 as '1' and '2' and four remaining sections. Cluster 1 comprises two British sections, Askrigg and Upper Bannock, and seven US sections, Allegheny River, Athens, Bonner Springs, Francis, Miller and Valencia. All nine contain limestones and only four, namely Upper Bannock, Allegheny River, Athens and Chelyan, contain vegetation beds.

Cluster 2 comprises nine British sections, Dosthill, Foggermount, Ingleton, Joppa, Neepsend, Overseal, Seaton Sluice and Sections A1 and A2. All these sections lack limestones and contain vegetation beds. Of the remaining four sections Blue Mound is distinct as it remains ungrouped even at a level at which all the other sections have joined into a single cluster. The other three sections, Dunbar, Columbia and Mulberry come from both sides of the Atlantic and are grouped together into a rather loose cluster at a level of 0.64. All three contain limestones and vegetation beds but sandy beds are either rare or absent and there is some tendency for vegetation beds to succeed limestones.

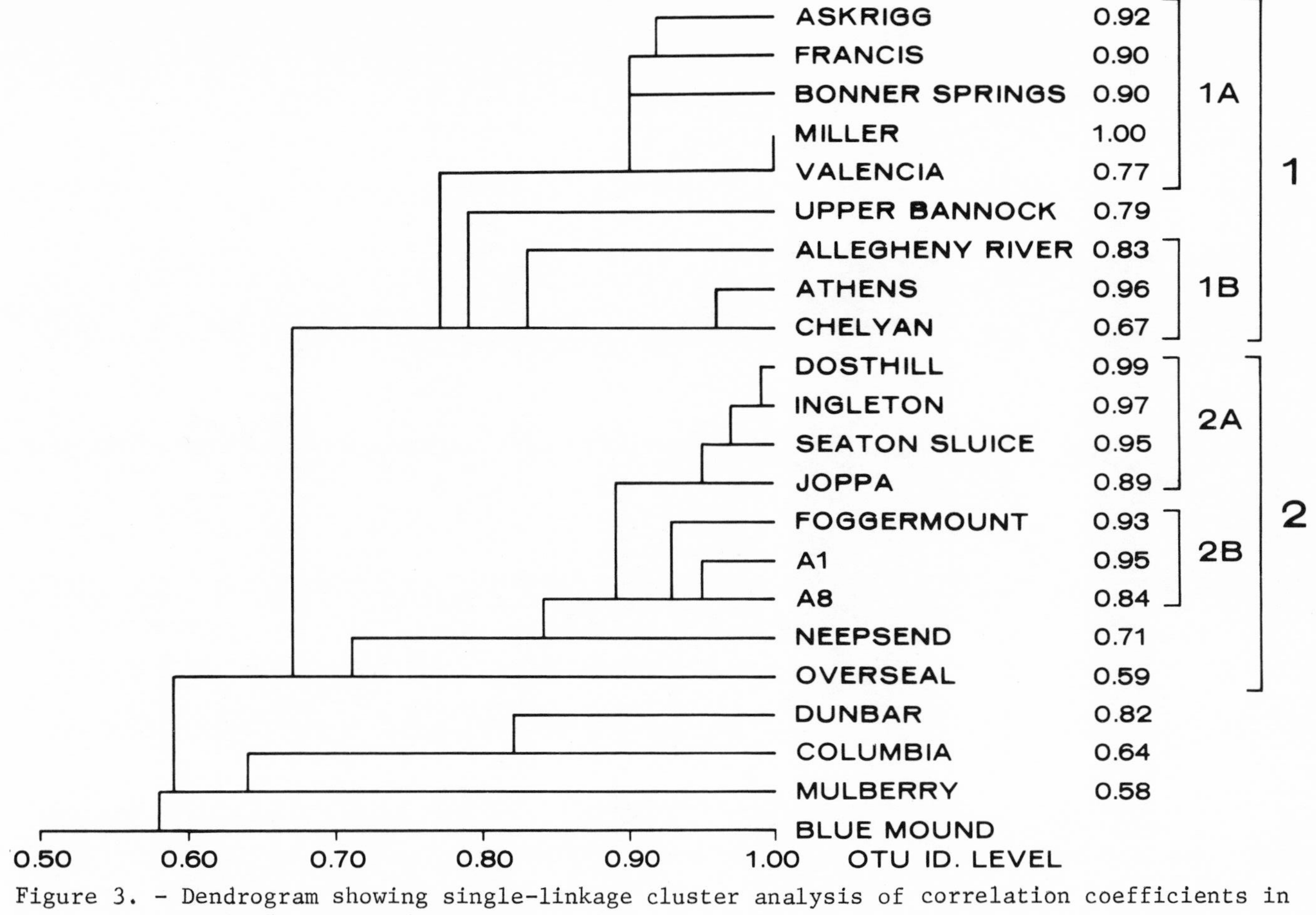

Figure 3. - Dendrogram showing single-linkage cluster analysis of correlation coefficients in Table 5. UK-US data set.

At the 0.80 level Group 1 splits into two clusters and one solitary section, Upper Bannock. The first cluster, which is marked 1A on Figure 3, comprises Askrigg, Bonner Springs, Francis, Miller and Valencia, all of which lack vegetation beds; the second cluster, marked as 1B, comprises Allegheny River, Athens and Chelyan. These sections contain relatively large numbers of vegetation beds, most of which are overlain by mudstone, and also a few limestones. The Upper Bannock section groups itself with Cluster 1B at a level of 0.79. This section contains both limestones and vegetation beds but the latter are overlain by sandy beds as commonly as they are by mudstone.

Cluster 2 remains intact at the 0.90 level except for Overseal which differs from the other sections in this cluster in the following respects: all sandy beds are capped by vegetation beds and there is an unusually high probability of mudstone being succeeded by vegetation beds.

Of the sections outside Clusters 1 and 2 only Dunbar and Columbia remain clustered together at the 0.80 level.

At the 0.90 level Cluster 1A remains intact. Within this cluster Miller and Valencia are grouped together at a level of 1.00 as it may be seen from Table 4 that the corresponding transition probabilities for these two sections are almost identical. In Cluster 1B Allegheny River splits at a level of 0.83. In this section, unlike Athens and Chelyan which remain clustered together, the probability of sandy beds being capped by vegetation beds is fairly high.

In Cluster 2 Neepsend splits at the 0.84 level. Unlike the sections remaining in this cluster it combines a high probability of sand beds passing up into mudstone with an even probability that vegetation beds will be overlain by mudstone or by sandy beds. At the 0.90 level Cluster 2 splits into two subsidiary clusters, marked 2A and 2B on Figure 3. In Cluster 2A, which comprises Dosthill, Ingleton, Joppa and Seaton Sluice, there is a high probability of sandy beds passing up into mudstone and a low probability of their passing up into vegetation beds. In Cluster 2B, which comprises Foggermount and Sections A1 and A8, the reverse is true. Sections A1 and A8 may be seen to be closely clustered together in Figure 3 in contrast to their wide separation in Figures 1 and 2.

Figure 4 shows the dendrogram produced by unweighted average-linkage cluster analysis of the data in Table 5. The cophenetic correlation coefficient is 0.73 so that again the dendrogram may not provide a completely adequate representation of the correlation coefficient data.

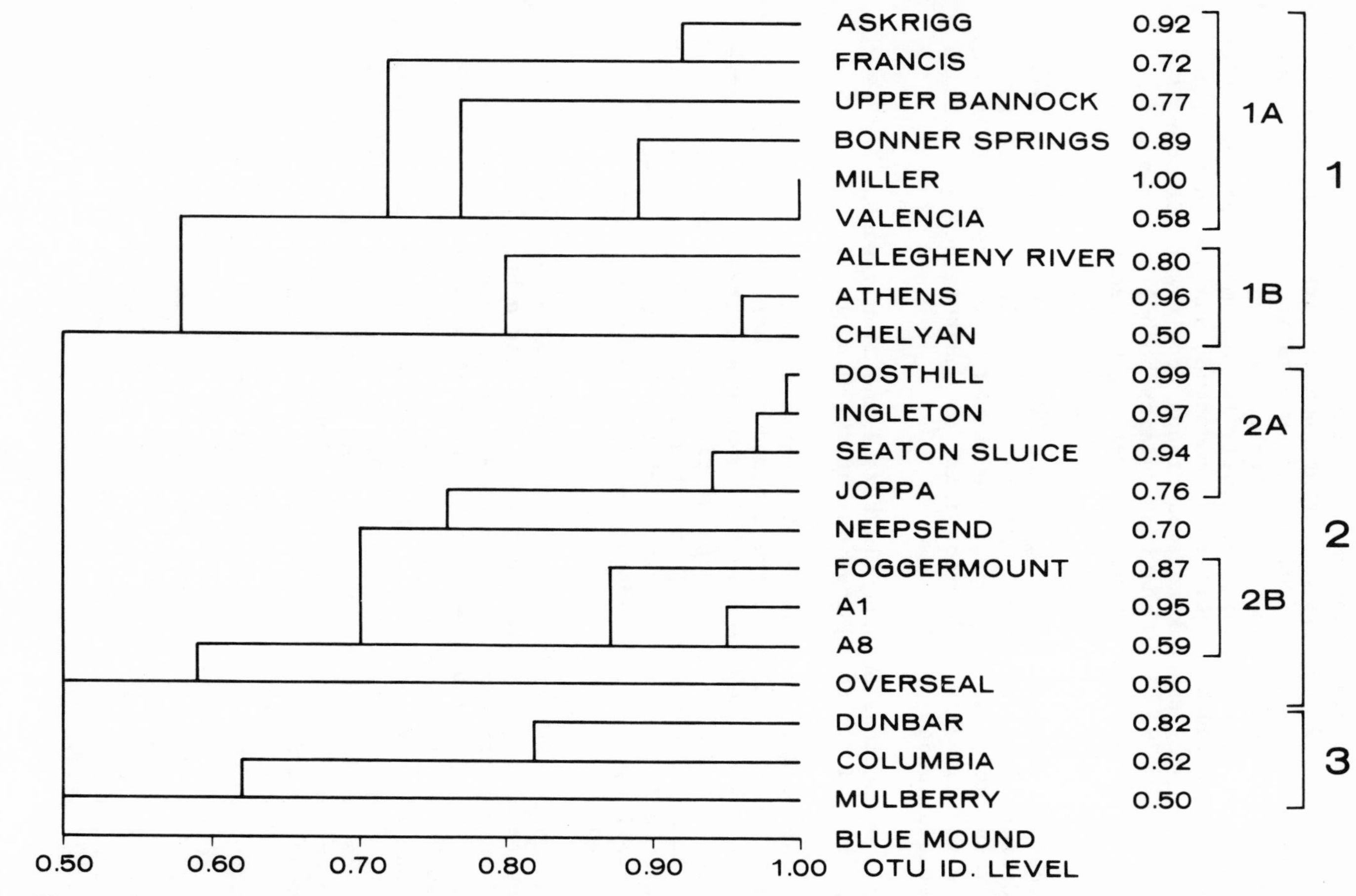

Figure 4. - Dendrogram showing unweighted average-linkage cluster analysis of correlation coefficients in Table 5. UK-US data set.

As might be expected, clustering takes place more gradually than in single-linkage cluster analysis and all 22 sections are grouped together at a level of 0.50. The clusters formed are almost identical to those in Figure 3 with the exception that Dunbar, Columbia and Mulberry form a loose but distinct cluster which has been marked '3' on Figure 4. The only other important differences are that Upper Bannock is included in Cluster 1A, although it contains sporadic vegetation beds, and Neepsend is loosely grouped with Cluster 2A.

SEDIMENTOLOGICAL INTERPRETATION

Because primary sedimentary structures and body- and trace-fossils have been adequately recorded in only a small proportion of the sections and because of the relatively small number of states recognized, it is possible to identify depositional environments in only the broadest terms, particularly in the heterogeneous UK-US data set. Nevertheless the results are of considerable interest and they indicate that the method may have considerable potential for distinguishing between various types of cyclical successions and for classifying these successions.

Kincardine Basin Data Set

The grouping of sections shown in the dendrograms, particularly the grouping obtained from unweighted average-linkage cluster analysis (Fig. 2), can readily be interpreted in terms of depositional environments. Thus the method leads to surprisingly clear results, despite the difficulties caused by close similarity of the transition probability values for many of the lithological transitions in all the sections within the data set and the relatively low cophenetic correlation coefficient values obtained for the dendrograms. The relative clarity of the results probably stems from the fact that such differences as are present in Table 2 tend to be associated with the critical transitions from mudstone to siltstone and from mudstone to sandstone. As will be explained later, both these transitions are of considerable sedimentological significance. Read (1965, 1969a, 1969b, 1970) has pointed out that the Limestone Coal Group of the Kincardine Basin closely resembles modern deltaic deposits and fits readily into the conceptual depositional model of deltaic cycles developed by Coleman and Gagliano (1964) and Frazier (1967, p. 288-291, figs. 2-3). An upward transition from mudstone directly into sandstone is indicative of the presence of upward-fining channel sandstones. These represent point-bar deposits and the coarser varieties of channel fill occupying erosive channels which have cut down through earlier distributary mouth bar sands (Fisk, 1955, p. 388; 1960,

p. 163; Welder in Fisher and others, 1969, fig. 13). By contrast a gradational upward passage from mudstone into siltstone and from siltstone into sandstone is indicative of the presence of upward-coarsening sheet sandstones. These are analogous to delta front silts and sands which grade upwards from prodelta clays and which are derived from sand and silt brought down by distributary channels and subsequently redistributed by waves and currents to form sheetlike bodies (Fisk, 1955, p. 386, 388; Frazier, 1967, figs. 4c, 7-10; Fisher and others, 1969, p. 24). In a differentially subsiding basin such as the Kincardine Basin, channel sandstones tend to be more prominent on the margins, where subsidence is low, rather than in the center of the basin. In addition, in the Kincardine Basin erosive channels carrying sand entered the area from the northwest and northeast and tended to occupy belts which extend southeastwards and southwestwards, respectively, across the basin (Read, 1961, fig. 8; Read and Dean, 1967, p. 151-153, figs. 3 and 4; Read, 1970). The section which differs most markedly from all the others is Section A1. It is located in the extreme northwestern part of the basin at the point where erosive channels from the northwest entered the basin and consequently where channel sandstones are most commonly represented in the sequence. The sections in Cluster X in Figure 2 either tend to lie close to the margins of the basin or else within the belts most commonly occupied by erosive channels carrying sand, whereas the sections in Cluster Y tend to lie nearer the center of the basin in areas more remote from these belts (Read and Dean, 1967, fig. 1)

The results obtained from the Kincardine Basin data set and their interpretation as outlined, suggest that the technique is able to extract useful sedimentological information from a remarkably uniform succession deposited in a single, small basin of deposition.

UK-US Data Set

The groupings of sections shown by the dendrograms in Figure 3 and 4 also are readily interpretable in terms of depositional environments, although the limitations and heterogeneous nature of the data necessarily lead to this interpretation being much less detailed than the interpretation given for the Kincardine Basin data set. Thus difficulties such as the small numbers of lithological members in some sections, the presence of some negative correlation coefficient values in Table 5 and the relatively low cophenetic correlation coefficients obtained for the dendrograms do not seem to have prevented the emergence of clear results.

The sections in Cluster 1 all represent environments which were not affected continuously by the deposition of wedges of clastic sediments associated with the building of delta lobes into the depositional basin. In the intervals between the formation of clastic wedges it was possible for limestones to be laid down, generally in a clearer water offshore environment. The sections grouped together in Cluster 1A, as shown in Figure 3, namely Askrigg, Bonner Springs, Francis, Miller and Valencia, were all deposited in a dominantly offshore environment, mostly in areas of relatively low subsidence underlain by tectonically stable basements. Moore (1958, p. 131) considers that the Askrigg succession is affected by periodic advances of deltas into a shallow sea and the same is probably true to some extent of the other sections. The delta lobes however were not built up to near sealevel and so were not colonized by land vegetation (Read, Dean, and Cole, 1971). Significantly Miller and Valencia, the two sections which are clustered most closely together, both come from the Wabaunsee Group (Virgilian) in eastern Kansas.

The Upper Bannock section, which is included in Cluster 1A in Figure 4 but excluded in Figure 3, differs from the four sections just described in that it was laid down near the margins of a subsiding volcanic land mass. In the lower part of this section clear water carbonate sedimentation was dominant but, later, delta lobes of sandy sediment appeared. Relatively few of these were able to fill the basin locally to near sealevel but some did so and were capped by vegetation. The 50-percent probability of these vegetation beds being overlain by sandy beds suggests the presence of channels which cut down into the underlying sediments (Read _in_ Francis and others, 1970, p. 168-178).

The sections in Cluster 1B all come from Pennsylvanian successions in the Appalachian Basin where deposition is known to have been strongly influenced by the building of delta lobes (Morris, 1967; Ferm and Cavaroc, 1969; Ferm, 1970; Ferm and others, 1970). In the Allegheny River section lobes of sandy sediment invaded the area of deposition more frequently and were more commonly colonized by vegetation than in the Athens and Chelyan sections.

Cluster 2 includes all the typical British Coal Measures facies sections. These were deposited in environments that were affected more or less continuously by the deposition of delta lobes or crevasse systems (Goodlet, 1959; Duff and Walton, 1962, p. 250-251; Elliott, 1968, 1969; Read, 1969b) so that the deposition of limestone is extremely rare. Cluster 2A comprises the Dosthill, Ingleton and Seaton Sluice sections, all of which are from the Westphalian Coal Measures of England, plus Joppa from the Westphalian of eastern Scotland. In all of these sections lobes of sandy sediment were somewhat less prominent than in the sections of Cluster 2B, and their tops were generally not colonized by land vegetation. All the sections in Cluster 2B come from Stirling in central Scotland. They lay nearer to the sources of coarse-grained clastic sediment than did the sections in Cluster 2A, in an area close to the points at which sand entered the Midland Valley of Scotland from the north. In this area the Namurian 'coal measures' facies represented by Sections A1 and A8 is repeated in the true Westphalian Coal Measures represented by Foggermount. Lobes of sandy sediment were commonly built into the basin, filling it to near sealevel, and these sandy lobes were generally capped by vegetation. In the Neepsend section colonization by land vegetation took place comparatively infrequently and it is possible that this section may represent an environment in a more distal part of a delta complex than most of the other English Coal Measures sections. In the Overseal section, which tends to remain apart from the other sections in Cluster 2, lobes of sandy sediment rarely entered the area and the succession tends to be repetition of medstones and vegetation beds. This section is known to include the prominent Mansfield Marine Band (Mitchell and Stubblefield, 1948, p. 19-21, table 2), indicating that part of the succession at least was deposited some distance offshore.

In the sections of Cluster 3, as in those of Cluster 1A, limestone was deposited in intervals between the active building of clastic wedges but, unlike the sections of Cluster 1A, clastic sediments periodically filled the basins to near sealevel, allowing vegetation to grow from time to time. The clastic wedges however tended to be composed of relatively fine-grained sediments and intervals of limestone deposition were comparatively rare, in contrast to the conditions prevailing in the sections in Cluster 1A. Underclay limestones are fairly common in the Columbia and Mulberry sections, both of which come from the Cherokee Group of central USA, and vegetation also grew directly upon limestone in the Dunbar section, which is a distinctly unusual feature in the Visean of Britain.

The Blue Mound section tends to remain separate from all the other sections. It belongs to the Douglas Group (Virgilian) of Kansas which contains a somewhat heterogeneous collection of facies types which may include offshore, deltaic and even possible fluvial deposits. The presence of sandy beds, probably the infillings of erosive channels, immediately above half the total number of vegetation beds is an unusual feature in a succession which also contains marine limestones.

POSSIBLE FUTURE DEVELOPMENTS

This paper describes only the first attempt to use the technique and the procedure could be modified in several ways. One alternative approach, described earlier, is to include only lithological transitions thought to be of considerable environmental significance or to exclude transitions which have similar probability values for all the sections examined. Another approach would be to weight environmentally significant transitions in some manner.

Similarity coefficients other than product-moment correlation coefficients, e.g. coefficients based on taxonomic distance, (Sokal and Sneath, 1963, p. 143-154; Harbaugh and Merriam, 1968, p. 160-165) could be used to compare the vectors of transition probabilities. Other methods of classifying this similarity coefficient data also could be tried. As mentioned earlier, a study is in progress to assess the relative advantages and disadvantages of the method outlined in this paper as compared to using cross-association similarity coefficients to determine the degree of similarity between coded sets of succession data. Experience with the Kincardine Basin sections and with the original 20 natural sections from the UK-US data tends to suggest that the present method yields results that are more readily interpretable from a geological standpoint.

CONCLUSIONS

A relatively simple technique has been developed for the quantitative comparison of cyclically deposited successions. The succession data were coded into a limited number of lithological states and a first-order transition probability matrix was constructed for each section. The elements in this matrix were rearranged to form a vector of transition probabilities and product-moment correlation coefficients were used to compare the vectors for each possible pair of sections. Single-linkage and unweighted average-linkage cluster analysis then were applied to the resulting matrix of correlation coefficients in order to reveal more clearly the relationships between sections.

The method has yielded valid sedimentological results when applied to two different sets of succession data. The first set, termed the Kincardine Basin data set, was derived from fifteen cored borehole sections. All boreholes were through exactly the same rather uniform paralic facies succession, namely the upper part of the Scottish Limestone Coal Group (Namurian, E_1) and all situated within the same small basin of deposition. The second data set, which was termed the UK-US data set, is more heterogeneous and has been coded in considerably less detail. It comprises coded data from twenty natural sections in cyclically deposited successions drawn from a varied selection of locations and stratigraphical intervals within the Carboniferous of Britain and the Pennsylvanian of the USA. The two most dissimilar sections from the Kincardine Basin data set were recoded and added to the UK-US data set. The importance of the nature and number of lithological states selected is demonstrated by the loose relationship between the coded versions of these two sections if the five-state form of coding is used in contrast to close similarity if the simplified form of coding is used.

ACKNOWLEDGMENTS

The research for this paper was accomplished and the paper was written while the senior author (WAR) was Visiting Research Scientist at the Geological Research Group of the Kansas Geological Survey, The University of Kansas. We gratefully acknowledge the financial support of the Kansas Survey and Leverhume Trust during execution of this research. The NTSYS programs used in this study were made available at The University of Kansas by F. J. Rohlf, J. Kishpaugh and R. Bartcher and were run on The University of Kansas GE 635 computer.

Tha authors are particularly grateful to Mr. G. S. Srivastava of the Kansas Geological Survey and to Prof. A. J. Cole of St. Andrews University for programming assistance and helpful advice. Thanks are also due to the following for reading the manuscript and making helpful suggestions for its improvement: Dr. J. C. Davis, Mr. J. M. Dean, Mr. T. R. M. Lawrie, Dr. T. V. Loudon, Dr. M. J. McCullagh and Dr. R. C. Selley.

REFERENCES

Allen, J. R. L., 1964, Sedimentation in the modern delta of the River Niger, West Africa, in Deltaic and shallow marine deposits: Elsevier Publ. Co., Amsterdam, p. 26-34.

Allen, J. R. L., 1965a, Fining-upwards cycles in alluvial successions: Liverpool Manchester Geol. Jour., v. 4, p. 229-246.

Allen, J. R. L., 1965b, Late Quartenary Niger Delta and adjacent areas: Am. Assoc. Petroleum Geologists Bull., v. 49, no. 5, p. 547-600.

Allen, J. R. L., 1970, Sediments of the Niger Delta: a summary and review, in Deltaic sedimentation modern and ancient: Soc. Econ. Paleontologists Mineralogists Sp. Publ., no. 15, p. 138-151.

Chayes, F., 1960, On correlation between variables of constant sum: Jour. Geophysical Res., v. 65, no. 12, p. 4185-4193.

Coleman, J. M., and Gagliano, S. M., 1964, Cyclic sedimentation in the Mississippi River deltaic plain: Gulf-Coast Assoc. Geol. Socs. Trans., v. 14, p. 67-80.

Coleman, J. M., and Gagliano, S. M., 1965, Sedimentary structures: Mississippi River deltaic plain, in Primary sedimentary structures and their hydrodynamic interpretation: Soc. Econ. Paleontologists Mineralogists Sp. Publ. No. 12, p. 133-148.

Doveton, J. H., 1971, An application of Markov chain analysis to the Ayrshire Coal Measures succession: Scottish Jour. Geology, v. 7, pt. 1, p. 11-27.

Duff, P. McL. D., Hallam, A., and Walton, E. K., 1967, Cyclical sedimentation: Elsevier Publ. Co., Amsterdam, 280 p.

Duff, P. McL. D., and Walton, E. K., 1962, Statistical basis for cyclothems: a quantitative study of the sedimentary succession in the East Pennine Coalfield: Sedimentology, v. 1, no. 4, p. 235-255.

Elliott, R. E., 1968, Facies, sedimentation successions and cyclothems in Productive Coal Measures in the East Midlands, Great Britain: Mercian Geologist, v. 2, p. 351-372.

Elliott, R. E., 1969, Deltaic processes and episodes: the interpretation of Productive Coal Measures occurring in the East Midlands, Great Britain: Mercian Geologist, v. 3, p. 111-135.

Ferm, J. C., 1970, Allegheny deltaic deposits, in Deltaic sedimentation, modern and ancient: Soc. Econ. Paleontologists Mineralogists Sp. Publ. No. 15, p. 246-255.

Ferm. J. C., and Cavaroc, V. V., 1968, A non-marine model for the Allegheny of West Virginia, in Late Paleozoic and Mesozoic continental sedimentation, northeastern North America: Geol. Soc. America Sp. Paper No. 106, p. 1-19.

Ferm. J. C., and Cavaroc, V. V., 1969, A field guide to Allegheny deltaic deposits in the upper Ohio Valley: Ohio Geol. Soc. and Pittsburgh Geol. Soc. Field Guide, 21 p.

Fisher, W. L., Brown, L. F., Scott, A. J., and McGowen, J. H., 1969, Delta systems in the exploration for oil and gas: Bur. Econ. Geol., Univ. Texas at Austin, 212 p.

Fisher, W. L., and McGowen, J. H., 1967, Depositional systems in the Wilcox Group of Texas and their relationships to occurrence of oil and gas: Gulf-Coast Assoc. Geol. Socs. Trans., v. 17, p. 105-125.

Fisher, W. L., and McGowen, J. H., 1969, Depositional systems in the Wilcox Group (Eocene) of Texas and their relationship to occurrence of oil and gas: Am. Assoc. Petroleum Geologists Bull., v. 53, no. 1, p. 30-54.

Fisk, H. N., 1955, Sand facies of recent Mississippi delta deposits: Proc. 4th World Petroleum Cong., Section 1/C, Paper 3, p. 377-398.

Fisk, H. N., 1960, Recent Mississippi River sedimentation and peat accumulation: Congres. Avanc. Etud. Stratigr. Carb., p. 187-199.

Fisk, H. N., McFarlan, E., Kolb, C. R., and Wilbert, L. J., 1954, The sedimentary framework of the modern Mississippi delta: Jour. Sed. Pet., v. 24, no. 1, p. 76-99.

Francis, E. H., Forsyth, I. H., Read, W. A., and Armstrong, A., 1970, The geology of the Stirling district: Geol. Surv. U.K., Mem., 357 p.

Frazier, D. E., 1967, Recent deltaic deposits of the Mississippi River: their development and chronology: Gulf-Coast Assoc. Geol. Socs. Trans., v. 17, p. 287-315.

Goodlet, G. A., 1959, Mid-Carboniferous sedimentation in the Midland Valley of Scotland: Edinburgh Geol. Soc. Trans., v. 17, pt. 3, p. 217-240.

Harbaugh, J. W., and Merriam, D. F., 1968, Computer applications in stratigraphic analysis: John Wiley & Sons, New York, 282 p.

Kanes, W. H., 1970, Facies and development of the Colorado River Delta in Texas: in Deltaic sedimentation modern and ancient: Soc. Econ. Paleontologists Mineralogists Sp. Publ. No. 15, p. 78-106.

Krumbein, W. C., 1962, Open and closed number systems in stratigraphic mapping: Am. Assoc. Petroleum Geologists Bull., v. 46, no. 12, p. 2229-2245.

Krumbein, W. C., 1967, FORTRAN IV computer programs for Markov chain experiments in geology: Kansas Geol. Survey Computer Contr. 13, 38 p.

Krumbein, W. C., and Scherer, W., 1970, Structuring of observational data for Markov and semi-Markov models in geology: Office Naval Res. Tech. Rep. 15, 59 p.

Merriam, D. F., 1970, Comparison of British and American Carboniferous cyclic rock sequences: Jour. Intern. Assoc. Math. Geology, v. 2, no. 3, p. 241-264.

Merriam, D. F., and Sneath, P. H. A., 1967, Comparison of cyclic rock sequences using cross-association, in Essays in paleontology and stratigraphy: Kansas Univ. Dept. Geol. Sp. Publ. 2, p. 521-538.

Mitchell, G. H., and Stubblefield, C. J., 1948, The geology of the Leicestershire and South Derbyshire Coalfield: Geol. Survey of Gt. Britain Wartime Pamphlet, No. 22, 2nd ed., 46 p.

Moore, D., 1958, The Yoredale Series of upper Wensleydale and adjacent parts of northwest Yorkshire: Yorkshire Geol. Soc. Proc., v. 31, pt. 2, no. 5, p. 91-148.

Moore, D., 1959, Role of deltas in the formation of some British Lower Carboniferous cyclothems: Jour. Geology, v. 67, no. 5, p. 522-539.

Morris, D. A., 1967, Lower Conemaugh (Pennsylvanian) depositional environments and paleogeography in the Appalachian Coal Basin: Unpubl. doctoral dissertation, Kansas Univ., 521 p.

Oomkens, E., 1967, Depositional sequences and sand distribution in a deltaic complex: Geologie Mijnb., v. 46, p. 265-278.

Oomkens, E., 1970, Depositional sequences and sand distribution in the postglacial Rhone Delta complex, in Deltaic sedimentation modern and ancient: Soc. Econ. Paleontologists Mineralogists Sp. Publ. No. 15, p. 198-212.

Parks, J. M., 1966, Cluster analysis applied to multivariate geologic problems: Jour. Geology, v. 74, no. 5, pt. 2, p. 703-715.

Potter, P. E., and Blakely, R. F., 1968, Random processes and lithological transitions: Jour. Geology, v. 76, no. 2, p. 154-170.

Read, W. A., 1961, Aberrant cyclic sedimentation in the Limestone Coal Group of the Stirling Coalfield: Edinburgh Geol. Soc. Trans., v. 18, p. 271-292.

Read, W. A., 1965, Shoreward facies changes and their relation to cyclical sedimentation in part of the Namurian east of Stirling, Scotland: Scottish Jour. Geology, v. 1, pt. 1, p. 69-92.

Read, W. A., 1969a, Fluviatile deposits in Namurian rocks of central Scotland: Geol. Mag., v. 106, p. 331-347.

Read, W. A., 1969b, Analysis and simulation of Namurian sediments in central Scotland using a Markov-process model: Jour. Intern. Assoc. Math. Geology, v. 1, no. 2, p. 199-219.

Read, W. A., 1970, Cyclically-deposited Namurian sediments east of Stirling, Scotland: Unpubl. doctoral dissertation, Univ. London, 273 p.

Read, W. A., and Dean, J. M., 1967, A quantitative study of a sequence of coal-bearing cycles in the Namurian of central Scotland, 1: Sedimentology, v. 9, no. 2, p. 137-156.

Read, W. A., and Dean, J. M., 1968, A quantitative study of a sequence of coal-bearing cycles in the Namurian of central Scotland, 2: Sedimentology, v. 10, no. 2, p. 121-136.

Read, W. A., Dean, J. M., and Cole, A. J., 1971, Some Namurian (E_2) paralic sediments in central Scotland: an investigation of depositional environment and facies changes using iterative-fit trend-surface analysis: Geol. Soc. London Jour., v. 127, pt. 2, p. 137-176.

Read, W. A., and Sackin, M. J., in press, A quantitative comparison using cross-association of vertical sections of Namurian (E_1) paralic deposits in the Kincardine Basin, east of Stirling, Scotland: Inst. Geol. Sci., Rept. No. 71/14, 21 p.

Robertson, T., 1948, Rhythm in sedimentation and its interpretation: with particular reference to the Carboniferous sequences: Edinburgh Geol. Soc. Trans., v. 14, p. 141-175.

Sackin, M. J., Sneath, P. H. A., and Merriam, D. F., 1965, ALGOL program for cross-association of nonnumeric sequences using a medium-size computer: Kansas Geol. Survey Sp. Dist. Publ. 23, 36 p.

Schwarzacher, W., 1967, Some experiments to simulate the Pennsylvanian rock sequence of Kansas: Kansas Geol. Survey Computer Contr. 18, p. 5-14.

Schwarzacher, W., 1969, The use of Markov chains in the study of sedimentary cycles: Jour. Intern. Assoc. Math. Geology, v. 1, no. 1, p. 17-39.

Selley, R. C., 1970, Studies of sequence in sediments using a simple mathematical device: Geol. Soc. London Quart. Jour., v. 125, p. 557-575.

Sokal, R. R., and Sneath, P. H. A., 1963, Principles of numerical taxonomy: W.H. Freeman and Co., San Francisco, 359 p.

Wanless, H. R., and others, 1970, Late Paleozoic deltas in the central and eastern United States, in Deltaic sedimentation modern and ancient: Soc. Econ. Paleontologists Mineralogists Sp. Publ., No. 15, p. 215-245.

Weller, J. M., 1958, Cyclothems and larger sedimentary cycles of the Pennsylvanian: Jour. Geology, v. 66, no. 2, p. 195-207.

Williams, E. G., and Ferm, J. C., 1964, Sedimentary facies in the Lower Allegheny rocks of western Pennsylvania: Jour. Sed. Pet., v. 34, no. 3, p. 610-614.

MODELS FOR STUDYING THE OCCURRENCE OF LEAD AND ZINC IN A DELTAIC ENVIRONMENT

Richard A. Reyment

Uppsala Universitet

ABSTRACT

The relationships between the Pb and Zn content of sediments of the Niger Delta, West Africa, and Eh, organic content of the sediments, content of carbonate shell substance, Mn, P, S, and depth of origin of sample have been studied by the multivatiate statistical method of canonical correlation. The first model weighs the "predictor" variables of organic content and depth, in negative association, against a "response" vector dominated by P, negatively associated with Pb, Mn and S. The correlation of this vector with the original variables brings out the underlying relationship (Mn, -P, 2S). A second statistically significant, and almost equally important relationship for these data present the correlation between organic content and depth, in positive association, bound to a response vector dominated by Pb, Mn and P. The underlying relationship is shown to be between all elements in relation to depth and content of organic substance. The distribution of Pb and Zn is described by this vectorial association. The predictor and response structures of this root seem to express the distribution of the organic sedimentary component and the chemical constituents. A third, although nonsignificant, interrelationship weighs shells, with their soft parts, against Pb, negatively associated with P and S. Eh is of no account in the first two models but plays a certain role for the third root. A fourth set of nonsignificant canonical variates may represent random variation in the Eh determinations. Although the bivariate correlation coefficient between Pb and Zn is high, there is a difference in part of their patterns of distribution in our material.

INTRODUCTION

As part of an actuopalaoecological survey of the Niger Delta (see Reyment, 1969), the contents of several metals in samples of surface sediment, including lead and zinc, were determined by Dr. Lily Gustafsson of the Department of Analytical Chemistry, University of Uppsala. The graphical analyses of lead and zinc distribution in relation to depth, and thus distance from the shore, shows that the concentration of Pb in the surface sediment tends to be moderately well correlated with depth (r = 0.4). Zinc is less regularly distributed than lead and in only one transect could a graphical appraisal of the data disclose a definite correlation between depth and the concentration of this metal in the surface sediment. The correlation coefficient between Zn and depth is not significant (r = 0.27).

The highest concentrations for both metals were found at the same station (Pb = 49 ppm, Zn = 360 ppm). The host sediment at this station is rich in shell debris and shells of foraminifers, mainly globigerinids (many species of which contain relatively large amounts of zinc). The basic statistics for the two elements are presented in Table 1.

Table 1. Basic statistics for lead and zinc (n = 23)

	Range	Mean	St. dev.	Approximate content in seawater
Lead	14-49	37.4	7.30	< 0.005
Zinc	37-360	121.4	71.27	< 0.01
	Values in ppm			

Several variables were determined in both the sedimentary water and the supernatant water at the time of sampling. The pH and Eh of the interstitial and supernatant water were determined by standard methods, as well as the organic content of the sediment (OC), and content of calcium carbonate (CA). Microscope study of the sediments showed that CA is mainly related to shells and shell debris in the sediment. Other chemical components considered important in charting the distribution of Pb and Zn are: Mn and S and P from all sources. Depth is also an important variable.

Several avenues of quantitative approach are open to the investigator for studying the relationships between the microelements and physical factors of the environment. Perhaps the most readily accessible multivariate statistical procedure is that of principal components analysis or its surrogate, factor analysis, at least in its commonly practiced version. The PCA model is the most suitable if there is complete uncertainty as to which of the variables are "predictors" and which of them are "responses"; that is, whether some form of generalized regression relationship exists between the variables. Where there is an indication of a generalized regression structure, the method of canonical correlation offers an intuitively reasonable model. This procedure determines the correlations between sets of variables, where there are as many canonical correlations as elements in the smaller of the sets. A discussion of the interpretation of canonical correlations and attendant hazards is given in Blackith and Reyment (1971). The variables pH, Eh, OC, CA and D (depth) belong logically within the same set. It is obvious likewise that Pb and Zn must belong to the same set. The variables P and S are of a boundary nature. After analysis of a set of preliminary computations, in which different combinations of variables were tried in the two sets, I decided to group P and S together with Pb and Zn. Manganese was included in the study on an exploratory basis. Details of the sampling program and other features are given in Reyment (1969).

To increase the readability of the ensuing text I have used a simplified mode of expression when referring to the variables. Thus, I say 'Pb decreases' and not 'the content of Pb decreases', realizing full well that this will aggravate the pedant.

THE METHOD OF CANONICAL CORRELATION

The models for the study of the correlations between sets of variables were derived by the method of canonical correlation. The rationale for the choice of this procedure is that in an ecological-sedimentological system, some variables will function as predictors, and others will be dependent on these predictors; i.e., the other variables will function as responses. Although the statistical method of canonical correlations is among the older weapons in the armory of multivariate analysis, it is one that is seldom seen in use in the disciplines of which multivariate procedures form the backbone, and it seems to have been ignored almost entirely by biologists and geologists alike. I think that this is a reflection of the difficulty attached to interpretation of the results of such an analysis. For, in contrast with all other basic multivariate methods, the interpretation of canonical correlation studies requires that one have a more intimate knowledge of the data and sound a priori theories on which to work. In biology and geology, at least, the technique may be expected to find its most realistic applications in experimental projects (e.g. see Lee, 1969a; Lee and Middleton, 1967; Reyment and Ramden, 1970).

A useful manner of thinking about the underlying concepts of canonical correlation is in terms of its logical relationship with multiple regression.

The general multiple regression equation

$$y = \beta_0 + \beta_1 X_1 + \beta_2 X_2 + \beta_3 X_3 + \ldots\ldots\ \beta_p X_p \tag{1}$$

expresses the relationship between a response variable y (to be estimated) and the predictors X_i. Here, the β_i are the multiple regression coefficients and p denotes the number of variables. There are observations on each of the predictor variables $X_1, \ldots$ $\ldots, X_p$ (which are not necessarily random variables). The random variable y is dependent on the set of X's.

Canonical correlation is a natural extension of multiple regression. Here there are two sets of variates, say z_1 and z_2, which are mutually dependent; the calculations are directed towards finding the correlations between the sets and in reducing these relationships to the simplest form. Canonical correlation has an intuitive appeal as a model for cause-and-effect studies, with one set of variables as predictors and the other as responses.

If there are p_1 elements in predictor vector z_1 and p_2 elements in response vector z_2, and $p = p_1 + p_2$, and the matrix of correlations between all variables is partitioned in the following manner

$$\mathbf{R} = \left[\begin{array}{c|c} \mathbf{R}_{11} & \mathbf{R}_{12} \\ \hline \mathbf{R}_{21} & \mathbf{R}_{22} \end{array}\right].$$

The submatrix $\mathbf{R}_{11}$, of order p_1, contains the correlations between the elements of vector z_1, the submatrix $\mathbf{R}_{22}$, of order p_2, contains the correlations between the elements of vector z_2, and submatrices $\mathbf{R}_{12}$ and $\mathbf{R}_{21}$, of order $p_1 \times p_2$, contain the cross-correlations between the elements of the two subvectors.

The computations of canonical correlations are of the common eigenform of a large portion of multivariate analysis. Thus,

$$(\mathbf{R}_{22}^{-1}\mathbf{R}_{21}\mathbf{R}_{11}^{-1}\mathbf{R}_{12} - R_{c_i}\mathbf{I})\mathbf{d}_i = 0, \tag{2}$$

with the restriction,

$$\mathbf{d}_i'\mathbf{R}_{22}\mathbf{d}_i = 1. \qquad (p_2 \leq p_1)$$

Here, the vector $\mathbf{d}_i$ contains the coefficients relating to vector $\mathbf{z}_2$, the so-called 'right-hand set' for the i-th canonical correlation. The corresponding coefficients for vector $\mathbf{z}_1$, $\mathbf{c}_i$ are

$$\mathbf{c}_i = (\mathbf{R}_{11}{}^{-1}\mathbf{R}_{12}\mathbf{d}_i)/\sqrt{R_{c_i}} \quad . \tag{3}$$

These subsequent components are uncorrelated with succeeding counterparts. The i-th canonical correlation then may be interpreted as being the correlation between the transformed variates x and y, where

$$x_i = \mathbf{c}_i'\mathbf{z}_1, \text{ and}$$

$$y_i = \mathbf{d}_i'\mathbf{z}_2.$$

A useful backcheck when developing a computer program for canonical correlation analysis is to work out the bivariate correlation coefficients between x_i and y_i,

$$R_{c_i} = \frac{1}{N} \sum_{i=1}^{N} x_i y_i,$$

and see whether this agrees with the values of the canonical correlations originally obtained. The graph of the x_i against the y_i constitutes an important part of the analysis in ecological and sedimentological work.

The relative contribution of the elements of $\mathbf{c}_i$ and $\mathbf{d}_i$ to the i-th canonical correlation then has to be ascertained; this forms the most important aspect of the analysis. The usual method of attempting an appraisal of the contributions of the variables to a particular correlation by inspection of the elements of the predictor and response eigenvectors is approximate and not infrequently misleading. The most reliable approach is by calculating the correlations between the original variables and the derived canonical variates. These correlations are ascertained by doing the multiplications

$$\mathbf{g}_1 = \mathbf{R}_{11}\mathbf{c}, \text{ and}$$

$$\mathbf{g}_2 = \mathbf{R}_{22}\mathbf{d}. \tag{4}$$

One may think of the successive canonical correlations as extracting variance from the data, a terminology that survives from the early days of factor analysis. The proportion of variance extracted from the first set by the canonical component x is then $\mathbf{g}_1'\mathbf{g}_1/p_1$. Psychometricians have recently made use of the idea of analyzing the efficiency of the two sets chosen for canonical analysis (Love and Stewart, 1968). This work is clearly a step in the right direction, for in some studies, including the present one, it is difficult to decide on the most informative location of some variables - whether they are really predictors, or responses. The redundancy R_{dx} of set 1, given the presence of set 2, is

$$R_{dx} = \frac{\mathbf{g}_1'\mathbf{g}_1 R_c^2}{p_1}, \text{ and conversely,}$$
$$R_{dy} = \frac{\mathbf{g}_2'\mathbf{g}_2 R_c^2}{p_2}. \tag{5}$$

This measure expresses the amount of overlap between the two sets contained in, say, the first canonical relationship, viewed with respect to $\mathbf{z}_1$ added to available $\mathbf{z}_2$. In other words, the proportion of the variance in the first subvector which is redundant to the variance in available $\mathbf{z}_2$. (The shared variance between x and y is R_c^2.) The sums of the two redundancy measures (5), the total redundancy, express the proportion of the variance of set 1 that is redundant to the set 2 variance, and vice versa for the countering measure of total redundancy. For the first situation,

$$R_{d1} = \sum_{k=1}^{w} R_{dx_k}, \tag{6}$$

where w is the number of significant eigenvalues of the canonical equation.

I wrote a computer program for doing the foregoing computations and other studies. It computes the p_2 canonical correlation coefficients, the $p_1 \times w$ matrix of elements for subvector $\mathbf{z}_1$, the $p_2 \times w$ matrix for the subvector $\mathbf{z}_2$, the tests of significance for successive eigenvalues, the matrix of structure coefficients for both sets, $\mathbf{G}_1$ and $\mathbf{G}_2$, the values of R_{dx} and R_{dy}, their sums, the transformed values of x and y, and their graphs.

THE VARIABLES

A series of tests disclosed that pH of the interstitial water of the sediments contributes little information to the solution of the problem and this variable was therefore excluded. Slight diagnostic value is attributable to Eh, but its contribution was just of the order of magnitude as to warrant inclusion. The organic contents of the sediments (OC) represents not only organic substance incorporated in the sediment as an integrated constituent, but also the soft parts of organisms living at the moment of collection of the samples. The carbonate substance (CA) is likewise of primary and secondary origins; it consists almost entirely of calcium carbonate in both forms, aragonite and calcite, as shown by x-ray analysis of the samples. The observations for P and S represent total determinations of these elements in the sediments, no attempt having been made to differentiate between sources.

BIVARIATE CORRELATION

It is instructive to examine the significant bivariate correlations between the variables (Table 2). The high correlations between OC, and S, and S and Mn suggest reactions between decomposing organic substance and environmental components. The high correlation between D and P confirms a well-known fact (Degens, 1968, p. 117), that the content of phosphatic substances in a sediment in the nearshore zone tends to increase with distance from the coast (although only to a certain limit), Manganese and sulfur are strongly correlated as are also Pb and Zn, and Zn and P. The basic statistics for lead and zinc are listed in Table 1.

Table 2. Statistically significant bivariate correlations

Between	r_{ij}	Between	r_{ij}	Between	r_{ij}	Between	r_{ij}
OC, Pb	0.43	Pb, Zn	0.70	OC, Mn	0.67	Pb, P	0.76
OC, S	0.76	Pb, S	0.42	D, P	0.73	Zn, P	0.57
Mn, S	0.60						

CANONICAL CORRELATION ANALYSIS

The first canonical correlation is dominated by the following combinations of variables in the two canonical eigenvectors

$$(OC, -1.2D) \text{ and } (Pb, -2.6P, 1.4S).$$

This relationship suggests that where organic matter is negatively correlated with depth (here, distance from the shore), lead and sulfur tend to occur in positive association and in negative association with total phosphorous in the sediment. Mere inspection of the elements of the canonical eigenvectors is insufficient and the correlations between the original variables and the transformed variates are needed (Table 3). The canonical correlation associated with the vectors is 0.91, which is a high value for any type of data. The predictor set constitutes a canonical component in the environment showing that shell material and content of P increase with distance from the shore (see remarks on bivariate correlations). Organic substance, Mn and S are concentrated to the nearshore zone.

Table 3. - Correlations between original variables and derived canonical variates with redundancy measures.

	Predictors	R_{dx}	Responses	R_{dy}
$R_{c_1} = 0.91$	0.65 organic matter -0.45 $CaCO_3$ -0.78 depth	0.26	0.40 Mn -0.41 P 0.73 S	0.15
$R_{c_2} = 0.84$	0.63 organic matter -0.29 $CaCO_3$ 0.63 depth	0.15	0.81 Pb 0.41 Zn 0.76 Mn 0.84 P 0.43 S	0.32
$R_{c_3} = 0.53$ N.S.	-0.42 organic matter -0.81 $CaCO_3$	0.06	0.32 Zn -0.48 S	0.03

The second canonical correlation is 0.84, which also is highly significant. The canonical eigenvectors are dominated by the weightings

$$(2OC, -CA, 2D) \text{ and } (2Pb, 2Mn, P).$$

The vectors of correlations between the original variables and transformed variates show that the predictor structure is opposite to that of the first canonical correlation. This vector also is correlated with distance from the shore, but it represents OC bound to the sediment. The corresponding response set shows strong contributions from all variables, but with Pb, Mn and P occupying a position in the foreground. This canonical variate describes the modes of occurrence of most of the lead and zinc in the sediments sampled. The results for the patterns of distribution of lead and zinc confirm the impressions yielded by the preliminary graphical study referred to in the introduction.

The redundancy values, R_{dx} and R_{dy} of Table 3 show that 26 percent of set 1 variance is covered by the first component of set 2, and 15 percent of set 2 variance is covered by the first component of set 1. The second component of set 2 explains 15 percent of the variance of set 1, and 32 percent of the variance of set 2 is covered by the second component of set 1. The totals for the 2 suites of canonical variates are therefore roughly the same with R_{d1} = 41 percent and R_{d2} = 46 percent.

Although the third canonical correlation is not statistically significant (Table 4), it is instructive to consider it briefly. The predictor eigenvector is dominated by OC and CA, in positive association, thus representing soft parts and shell of organisms; Eh makes a noticeable contribution to this set. The response set is dominated by the variables Pb, P, and S in equal weightings. The lead is probably associated with living organisms with shells. It is possible that the third canonical correlation is a reflection of the subordinate position occupied by living organisms in the sediment relative to the accumulated dead material.

Table 4. - Significance of canonical correlations.

	R_{c_1}	Chi-square	Degrees of freedom
1	0.91	60.2	20
2	0.84	28.0	12
3	0.53	6.5	6
4	0.17	0.5	2

GRAPHICAL ANALYSIS

A useful adjunct to the analysis of the elements of the predictor and response sets is provided by plots of the transformed scores of the observational vectors, suitably partitioned. Figure 1 illustrates the plot of the scores for the first canonical correlation. As a result of the high correlation between the sets, all points group closely within an ellipsoidal field. The most significant point arising from this diagram is the clear grouping of the points into the transect stations which were sited in the nearshore zone and a second group of transect stations which were located in the outer reaches of the delta. The first canonical correlation thus reflects a depth-controlled relationship for the distribution of the elements.

The graph of the scores for the second canonical correlation

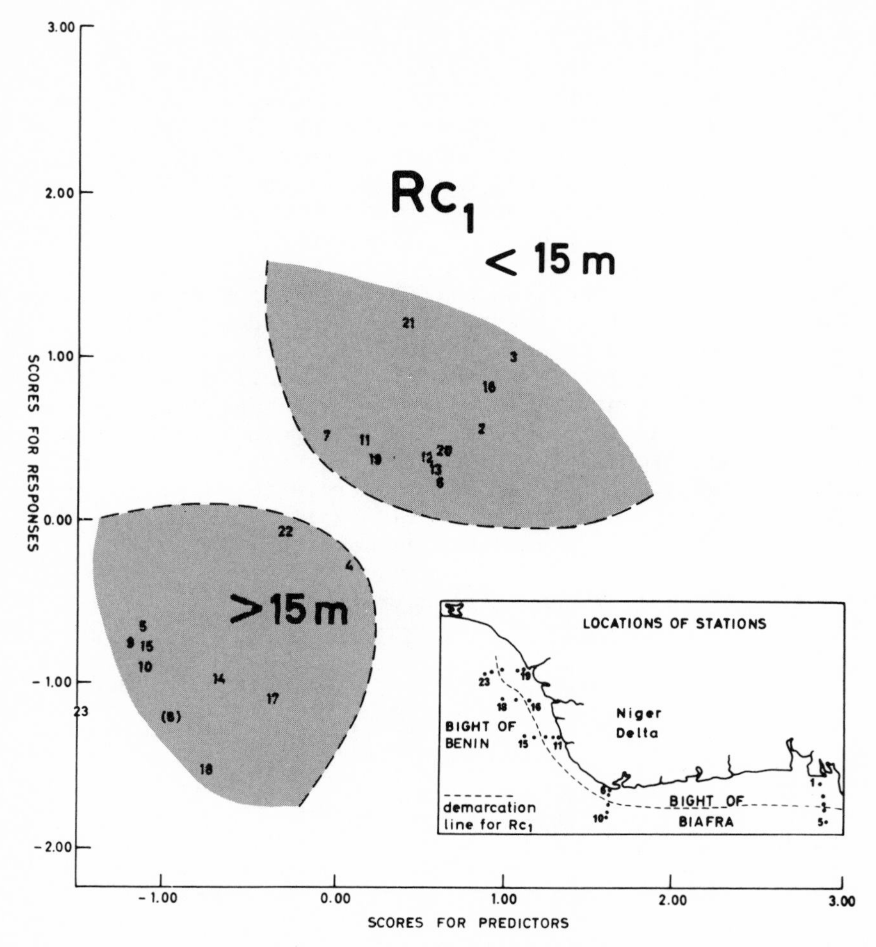

Figure 1. - Plot of canonical transformed scores for first canonical correlation. Inset map displays locations of stations analyzed.

is shown in Figure 2. No such clear pattern emerges as for the first pair of canonical variates, and this is not an unexpected result. Despite the rather small size of the sample, the fact that the transformed variates are right-skewed bivariate normal is evident. In general, the transect stations are grouped in a more complicated pattern than the first pairs, in which the interplay of all the elements studied, on the one hand, is balanced against the correlation of OC with depth. There is a slight tendency for some of the transect stations to cluster. This is marked on the figure.

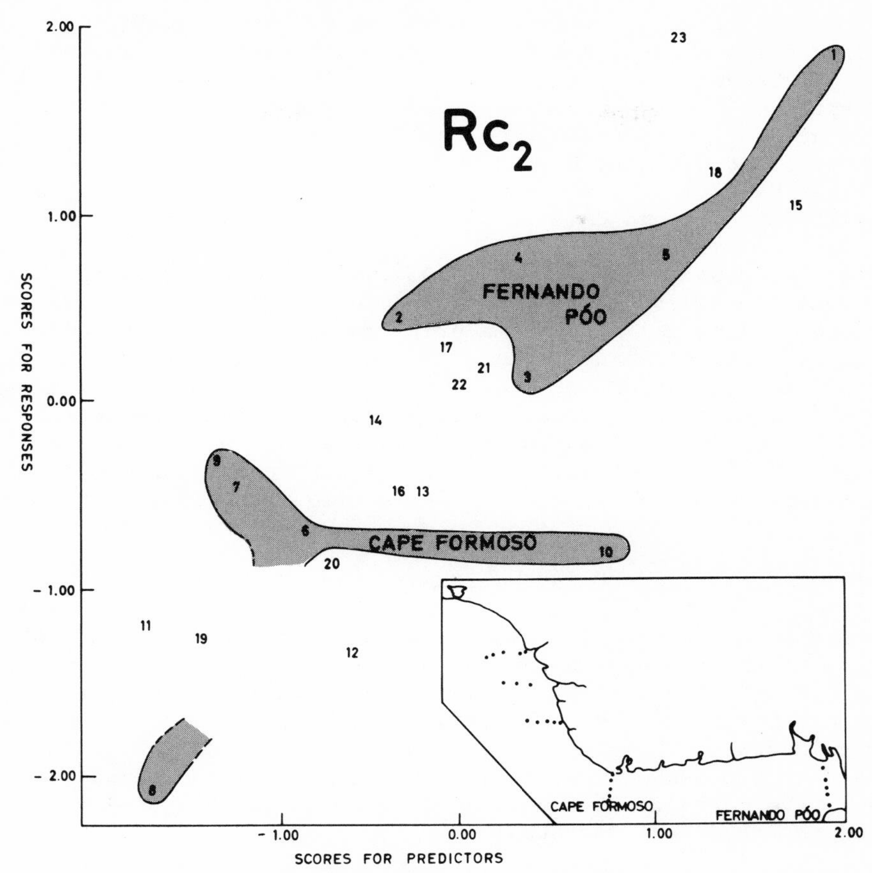

Figure 2. - Plot of transformed scores for second canonical correlation. Inset map indicates locations of two transects of particular interest.

PRINCIPAL COMPONENTS ANALYSIS

The most logical competitor to the method of canonical correlation for analyzing the data of the present problem is the technique of principal components analysis. The first component, which accounts for 36.7 percent of the total variation, is almost entirely composed of positive, almost equal contributions from the variables OC, Pb, Zn, P, and S. Thus there is a considerable difference here compared with the results yielded by the canonical eigenvectors, but reasonable agreement with the correlations between the original variables and the canonical variates for the second root. The second component, accounting for 23.5 percent of the total variation, unites OC, D, Zn, and P, on the one hand, and

Mn and S on the other, in negative association, which again is only partly comparable with the canonical analysis; Zn is an important constituent also here. The third principal component (13.9 percent of the total variation) is constructed of Eh and Zn in strong positive association, these variables being negatively correlated with CA, D, and Mn. The fourth principal component (11.0 percent of the total variation) is entirely explained by covariation in Eh and CA.

Hence the results of the principal components analysis of the data are not nearly so clear-cut as those of the canonical correlation study. I think that the material considered in this paper brings out a fundamental difference in the models underlying the two methods and one which may be imperfectly understood. This is that principal components should only be applied to homogeneous data in which the variables are of the same type. Where a cause-and-effect relationship exists, a multivariate regression model, such as canonical correlation, provides a more appropriate approach.

CONCLUDING REMARK

It has been shown how models for studying relationships existing in a sedimentary environment can be generated by the method of canonical correlation. These models may be used in exploratory prediction studies and they are useful also as a basis for simulation experiments on the computer. Finally, the interpretation of canonical correlation structures may be considerably bettered by the use of the correlation coefficients between the original variables and the canonical variates.

REFERENCES

Blackith, R. A., and Reyment, R. A., 1971, Multivariate morphometrics: Academic Press, London, 412 p.

Degens, E. T., 1968, Geochemie der Sedimente: Ferdinand Enke Verlag, Stuttgart, 282 p.

Lee, P. J., 1969a, The theory and application of canonical trend surfaces: Jour. Geology, v. 77, no. 3, p. 303-318.

Lee, P. J., 1969b, FORTRAN IV programs for canonical correlation and canonical trend-surface analysis: Kansas Geol. Survey Computer Contr. 32, 46p.

Lee, P. J., and Middleton, G. V., 1967, Application of canonical correlation to trend analysis, in Computer applications in the earth sciences: Colloquium on trend analysis: Kansas Geol. Survey Computer Contr. 12, p. 19-21.

Love, W. A., and Stewart, D. K., 1968, Interpreting canonical correlations: theory and practice: Am. Inst. Res., Palo Alto, 75 p.

Reyment, R. A., 1969, Interstitial ecology of the Niger Delta: Bull. Geol. Inst. Univ. Upsala, v. 1, p. 121-159.

Reyment, R. A., and Ramden, H. A., 1970, FORTRAN IV program for canonical variates analysis for the CDC 3600 computer: Kansas Geol. Survey Computer Contr. 47, 40 p.

THE SEMI-MARKOV PROCESS AS A GENERAL SEDIMENTATION MODEL

Walther Schwarzacher

The Queen's University, Belfast

ABSTRACT

Sedimentation may be considered as a two-stage process. Environmental control determines the gross aspect of lithology but other sedimentation processes are responsible for bedding. The two are separated because they operate according to different time scales. The semi-Markov process provides an ideal model for this situation, under the condition that the environmental history is determined by a Markov chain.

Models of bed formation can be based on the theory of random walk. The resulting bed thickness is either exponentially or Gamma distributed. Gamma distributions seem to fit observed bed-thickness better than the lognormal distribution. The semi-Markov process may result in exponential or polymodal bed thickness distributions. Examples of both have been generated by simulation.

INTRODUCTION

The important question of bed-thickness distributions in stratigraphic sections was raised recently by Krumbein and Dacey (1969). They show that the now increasingly used Markov model leads to an exponential bed thickness of sediment deposited at discrete time intervals and the sequence of rock types is determined by the Markov chain P_{ij}. The thickness of a stratum thus is given by the number of steps for which the system remains in a given state. For example, the system will remain in state i for as long as it takes for this state to be succeeded by a different one. Therefore,

$$P(T_i = K) = P_{ii}^{k-1} P_{ij}, j \neq i \qquad (1)$$

is the probability of state i obtaining the thickness $T_i = k$, i.e. k 'thickness-steps' (Krumbein and Dacey, 1969, p. 84). If thickness distributions are observed which do not conform to this distribution, then this simple model is not applicable. Krumbein and Dacey (1969, p. 93) suggest that a semi-Markov process could be used to explain nonexponential thickness distributions. Specialized semi-Markov processes have been employed previously by Vistelius and Feigel'son (1965) and Schwarzacher (1968), who used the theory of Markov chains to derive their models. The semi-Markov chain as a model of stratigraphic successions was developed in some detail by Dacey and Krumbein (1970); this important paper was unfortunately not available to the author at the time of writing. However, by using renewal theory, some problems connected with semi-Markov processes can be solved more easily and it seems that the semi-Markov process can be used as a general and yet simple model of sedimentation. The purpose of this paper is to concentrate on this particular sedimentation model; at the same time, the semi-Markov process probably has many more other applications in geology.

A GENERAL SEDIMENTATION MODEL

A semi-Markov process may be visualized as follows. A system passes from state to state following the transition probabilities of a discrete Markov chain A_{ij}. The transitions however are not instantaneous but a certain time x is spent in the process of changing states. The time spent in transition is a random variable that is specified by a probability density function $f(x)_{ij}$; the subscript i, j indicates that each type of transition may have its own statistical properties. Figure 1 illustrates this process for a two-state system.

In dealing with transitions from i to j the interval leading to event j, is called by convention j-type interval and similarly

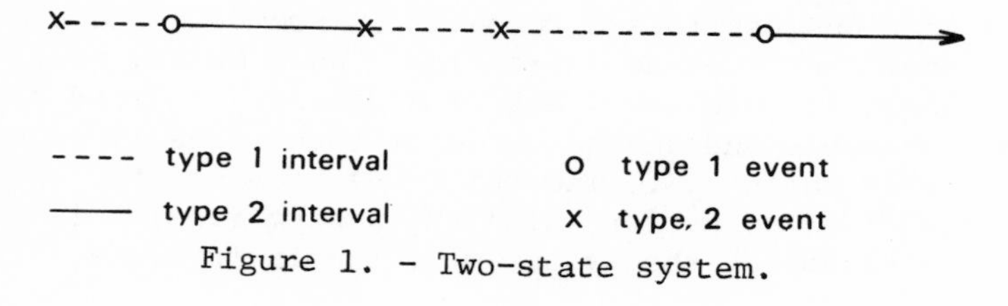

Figure 1. - Two-state system.

a transition terminated by i, is called an i-type interval. The matrix A represents an embedded Markov chain and this together with matrix $f(x)_{ij}$ is needed to calculate the transition matrix $P(t)_{ij}$ of the semi-Markov process. This transition matrix is a function of time, and $p(t)_{ij}$ stands for the probability that state j is entered at time t, provided the system has been in state i at time t_o. A compact formula for deriving $P(t)$ from A and $f(x)$ has been given by Cox and Miller (1965, p. 354).

Let

$$g^*(S) = \begin{bmatrix} a_{11}f_{11}^*(S) & a_{12}f^*_{12}(S)\ldots \\ a_{21}f_{21}^*(S) & \end{bmatrix},$$

where the a's are elements of A and an asterisk indicates the Laplace transform then it can be shown that

$$sP^*(s) = (I - g)^{-1}(A - g^*). \qquad (2)$$

The semi-Markov process can be adapted as a sedimentation model in the following manner. Consider the stratigraphic record as the result of a two-stage process. Firstly, environmental conditions determine the lithology; secondly, there exists a mechanism of sedimentation which determines the amount of sediment to be laid down. This can be expressed in the following scheme:

Environmental time History	→	Sedimentation Mechanism	→	Stratigraphic Record
A		$f(x)$		$P(t)$

It is assumed that the events in the environment are controlled by a discrete Markov chain. A transition takes place at regular time intervals, as it were, at each tick of a clock. (Krumbein uses the picturesque term of a "Markovian clock". There is, however, nothing Markovian in our clock mechanism which is supposed to tick at regular intervals.) Such transitions may either be from one state to itself or into a new state. By environmental states are meant a set of conditions which will lead to a definite lithology, for example in a two-state system one might designate a sand and shale state meaning that the environment in state one is suitable for the deposition of sand and in state two shale can be potentially deposited. The environmental history is a time series in the literal sense of the word, hence the postulate of the regular ticks. Matrix A being Markovian implied that the environmental history is a random process and in particular that the time periods which the system spends in a definite state are given by the exponential distribution equation (1).

The second assumption is that whenever a transition in the environmental history occurs, a certain thickness of sediment is laid down which, in geological terms, is expressed by saying a bed is formed. If the environmental history moves from a particular state back to the identical state two cases can be considered. Either each transition produces a recognizable bed meaning that a series of identical beds follow each other, but because of the semi-Markov property they will differ in thickness. Alternatively, one may specify that the beds are only recognizable when a new state is entered. The same situation applies in the simple Markov sedimentation model where the recurrence of identical states has been termed multistory lithologies (Carr and others, 1966) and in which the second case is recognized by a zero diagonal in the transition matrix. In the Markov model in which the environmental time history is directly identified with the produced lithological states, any transition matrix can be reduced to one with zero diagonals (Schwarzacher, 1969). In this instance such a matrix is referred to as an embedded Markov process (Krumbein and Dacey, 1969). This means that the matrix describes the sequence of lithologies in the stratigraphic series. In the semi-Markov model, the embedded chain is given by matrix A which generally does not have zero probabilities in the diagonal.

If each transition from one state to itself is recognizable, then the analysis of a semi-Markov process is relatively easy but complications arise if the transitions are not recorded. Unfortunately, the later case is more likely to occur in nature. To clarify the general principles of the semi-Markov sedimentation model, the following deliberately naive example may be used. Consider a small basin which receives two types of sediment sand when it rains and clay when it is fair. It is further assumed that rainy and fair days follow each other in a Markov sequence. Let each rainy day produce a bed of sandstone. The thickness of such beds is a random variable which is determined by some mechanism of bed formation. If it happens that this mechansim is known, then the model will provide a full description of the stratigraphic sequence. Two possibilities now have to be considered; either each rainy day produces a separate sandstone bed, or successive rainy days produce a single bed which results from merging individual beds. The two cases are examples of recognizable and nonrecognizable multistory lithologies. Possibly the most startling feature of the model is that although the environmental history has been related to time steps, the thickness of individual beds has not. At first this may seem unrealistic but in fact, it is one of the most useful aspects of the model. As will be shown later, it may be possible to separate the sedimentation process which works on a much more detailed time scale than the time events in environmental history.

MODELS OF BED FORMATION

Sedimentary bedding has several definitions but sedimentation conditions during the formation of a bed are assumed essentially constant (Otto, 1938, p. 575). For our purpose a bed is the smallest possible stratigraphic unit which in itself contains no recognizable bedding plane or break. Beds may range in thickness from laminae to units of considerable thickness. The simplest hypothesis of bed formation relates the thickness of a bed directly to the duration of the environmental regime, that is, while environmental conditions are constant, steady sedimentation continues. The sedimentation mechanism therefore is assumed to be deterministic with sedimentation providing a faithful record of the environmental time history.

The Exponential Distribution

It is more realistic to regard sedimentation as a noncontinuous process, in which random periods of deposition and erosion alternate. Each break, which may be either an interval of nondeposition or erosion, causes a bedding plane. This model was probably first suggested by Vistelius (1949) and Kolmogorov (1951). If the simplifying assumption is made, that the probability of deposition p, and the probability of erosion q, remain constant, then the growth of the sediment can be regarded as a simple random walk which either moves up or down in unit steps. The transition probability matrix for this random walk can be written

$$P = \begin{bmatrix} p & q \\ p & q \end{bmatrix},$$

the columns being identical because it may be assumed that erosion and deposition are independent variables. Once again, by referring to equation (1), the periods of deposition are found to be exponentially distributed. That the bed-thickness distribution should be exponential, is to be expected and this can be seen immediately if we regard the erosional stages purely as markers. If erosion occurs, the thickness distribution will be truncated but will remain exponential. The random walk model contains two essential elements, firstly, the individual steps (transitions in the matrix P) which by definition are instantaneous; secondly, intervals of continuous sedimentation which are separated by breaks and thus constitute beds. The elementary steps must be present in all sedimentation processes. They may be individual sand grains or even molecules involved in the formation of a crystal. Steps which are equivalent to sedimentary bedding are somewhat more subjective as they involve our ability to recognize breaks in sedimentation. Obviously, the accuracy of the observations will differ according to the methods which are used. For example, if bedding is measured from aerial photographs or measured under the microscope, phenomena of different intensities are observed. Allowance for this can be made in the model by specifying breaks of different intensity, for example, by separating breaks caused by one, two or more consecutive periods of nondeposition. It follows from the properties of the exponential distribution that higher order breaks are exponentially distributed.

The bed unit has, by definition, the property that it cannot be subdivided into time stages and even relative dating within it may be impossible. A bed therefore represents a discrete step to the geologist, which because it cannot be subdivided, is assumed to have formed instantaneously. Mathematically the formation of a bed can be treated as a point event. This is not to imply that bed formation is instantaneous in a physical sense. In fact two different time scales are being dealt with here, the scale of the stratigrapher being somewhat coarser than that of the physicist. Difficulties arise because in discussing bed formation, we enter the domain of the latter, whereas when discussing environmental history we usually employ geological scales. It follows that bed formation as caused by the random walk will operate independently of the environment and will occur while the environmental conditions are essentially constant. Hairsplitting can be avoided by accepting the environmental state as a statistical equilibrium. In this context the experimental work of Jopling (1964) is relevant. He has shown that laminations in artificially deposited sands develop under steady flow conditions and at fast rates. Similar fluctuations probably occur in all sedimentation processes. If one lets the elementary steps become small, the discrete distribution of equaiton (1) can be replaced with the continuous negative exponential distribution

$$f(x) = \rho e^{-\rho x}. \tag{3}$$

Here the parameter ρ is a density of events and is given by $1/\mu$ where μ is the mean of the distribution. If a steadily growing sediment is considered in which breaks occur with a given density but are otherwise completely random, then the spacings between the breaks will correspond to the distribution (3). The negative exponential, being a continuous distribution ,is more easily handled than (1) and is also more appropriate to the essential physical nature of bed formation.

The Gamma Distribution

The previous section has shown that accepting a probabilistic model of bed formation implies that the process becomes independent of the environment in the sense that it will proceed automatically as long as environmental conditions are suitable. Events in the environment follow each other at time intervals given by one scale. Bed formation is caused at intervals which are provided by a second time scale. There are two possible situations; either the intervals of bed formation are shorter than the environmental events, or they coincide. The third possibility that bed formation takes longer than a certain environmental state leads again to the deterministic concept in which the thickness of beds is strictly controlled by the duration of the environmental regime. The first possibility of several short fluctuations within an environmental time interval causes no conceptual difficulty and is indeed the most likely situation to be encountered. Applied to the sand-shale example, beds could be associated with individual rain showers and each rainy day therefore will contain several beds. The second situation may be obtained by decreasing the steps used for structuring the environmental history until they are so short to coincide with the duration of a shower. In this situation each shower triggers the random process of bed formation. Alternatively there may be some mechanism which determines that only one bed within a given time interval can be formed. Most geological processes seemingly have built-in thresholds which regulate sedimentation in reaction to the environment and more information on this is needed. To put it again as simply as possible, one raindrop will not cause sand sedimentation, a single shower or an average rainy day might well do so.

The sand-shale example now could be based on the following realistic assumptions. Rain showers produce exponentially distributed laminae, which are the fluctuations within the sedimentation of the sand regime. In this particular example it is unlikely that each day is marked by a separate bed, i.e. no multistory lithologies are recorded. A boundary will only occur at the end of a period of rainy days. In order to describe this process by the semi-Markov model, the thickness distribution of the amount of sediment laid down per day must be known. This is obviously the sum of the thicknesses laid down in each shower. For simplicity we can assume that each day has a constant number of showers k, each of which produces a layer. The thicknesses of these layers are independent random variables, $x_1, x_2, \ldots, x_k$, which are identically distributed as equation (3). Most textbooks show that the distribution of independent sums is given by the convolution of the distribution functions concerned. Specifically for exponential random variables, the Laplace transform is given by

$$f^*(x) = \frac{\rho}{\rho + s} \quad \text{and the convolution:}$$

$$f^*(x_1 + x_2 + \ldots x_k) = \frac{\rho^k}{(\rho + s)^k} .$$

This has the inverse

$$f(x) = \frac{\rho^k}{\Gamma(k)} x^{k-1} e^{-\rho x} . \tag{4}$$

Equation 4 is known as the Gamma distribution with the properties

$$E(x) = \mu = \frac{1}{\rho} , \text{ var}(x) = \frac{\mu^2}{k} , \frac{\sigma}{\mu} = \frac{1}{k} .$$

If $k = 1$ the distribution becomes the negative exponential distribution which, as can be seen from (3) and (4), is only a special situation of the Gamma distribution for $k > 1$, the distribution is zero at the origin and rises to a single maximum at $x = k - 1/\rho$ When k becomes large, the distribution is more and more symmetrical approaching the normal distribution.

The concept of thresholds in sedimentation processes makes it likely that most bedding consists in fact of several stages of deposition and the Gamma distribution is therefore the logical bed-thickness distribution which arises from the random walk model. As will be shown, Gamma distributions with relatively low k values fit empirical data extremely well.

SEMI-MARKOV MODEL IN A TWO-STATE SYSTEM

The sedimentation model now can be applied to a simple two-state system. As a geological illustration, the previous example of sand and shale sedimentation may be used. The following data are assumed known. Firstly, the time history (sequence of rainy days) as represented by matrix A which may be written as

$$A = \begin{bmatrix} a_{11} & a_{12} \\ a_{21} & a_{22} \end{bmatrix} .$$

Secondly, the process of bed formation is assumed known and in particular, the probability density function $f(x)_{ij}$, to which this process will lead. This is expressed by matrix

$$G^*_{(s)} = \begin{bmatrix} a_{11}f^*_{11}(s) & a_{11}f^*_{12}(s) \\ a_{21}f^*_{21}(s) & a_{22}f^*_{22}(s) \end{bmatrix},$$

in which the asterisk once more denotes the Laplace transform. Using equation (2), matrix $P_{ij}(t)$ may be calculated, which gives the transition probability matrix as a function of bed thickness. Inspection of equation (2) will show that specific solutions for higher order matrices or even for the two-state example, can become extremely cumbersome and in general, solutions by Monte Carlo methods will be far more profitable. However, in special situations, surprisingly simple and yet instructuve solutions can be obtained.

For example let

$$A = \begin{matrix} .5 & .5 \\ 1.0 & 0 \end{matrix}. \qquad (5)$$

This is a Markov chain in which as soon as state two is entered, state one will follow immediately. Let it be assumed further that the processes of bed formation are identical for all types of transitions, i.e. $f(x)_{ij} = f(x)$ then equation (2) can be simplified to

$$P^*(s) = \frac{1}{1 - .5f^*_{(s)} - .5f^*_{(s)}{}^2} \begin{bmatrix} .5(1 - f^*_{(s)}{}^2) & .5(1 - f^*(s) \\ 1 - f^*_{(s)} & .5f^*_{(s)}(1-f^*_{(s)}) \end{bmatrix}.$$

Assuming $f(x)$ to be the negative exponential distribution, its Laplace transform, $\frac{\rho}{(\rho+s)}$, can be substituted for $f^*(s)$ and the following is obtained

$$P^*(s) = \frac{1}{s(s + 1.5)} \begin{bmatrix} .5s + \rho & .55 + .5\rho \\ s + \rho & .5\rho \end{bmatrix}.$$

Setting $\rho = 1.0$ the backward transformation gives

$$P_{ij}(t) = \begin{bmatrix} .666 - .1666e^{-1.5t} & .1666^{-1.5t} + .3333 \\ .333e^{-1.5t} + .666 & .333 - .333e^{-1.5t} \end{bmatrix}. \qquad (6)$$

To understand fully the meaning of $P_{ij}(t)$ reference may be made to Figure 1. The leading element of the above matrix refers to transitions from state 1 to state 1 and $P_{11}(t)$ has the meaning: an event 1 has just occurred, i.e. a bed of type 1 is completed, the probability that it will be followed by another bed of type 1 with thickness t is given by the expression $P_{11}(t)$. It may be seen that if t in equation (6) approaches zero, i.e. if the transitions are instantaneous, the series of events becomes again the Markov chain of (5), which according to our meaning are time events without any stratigraphic thickness associated with it. On the other hand, when t becomes large and the exponential term approaches zero, the transition probabilities become identical with the stable probability vector of A,

$$\Pi = \frac{a_{21}}{a_{12} + a_{21}}, \quad \frac{a_{12}}{a_{12} + a_{21}} = .666, \ .333.$$

Let us next consider the probably more common situation in which multistory lithologies are not recognized, that is, transitions between identical states are not recorded. The effect is the same as if matrix A (equation (5)) had taken the form

$$A' = \begin{bmatrix} 0 & 1 \\ 1 & 0 \end{bmatrix}.$$

The densities however with which state one and two occur are unaltered and given by the stable vector Π of (5). We shall modify therefore the bed-formation process and write

$$f_1(x) = \rho_1 \Pi_1 e^{-\Pi_1 \rho_1 x} \quad \text{and} \quad f_2(x) = \rho_2 \Pi_2 e^{-\rho_2 \Pi_2 x}. \tag{7}$$

Equation (2) using A simplifies to

$$sP^*(s) = \frac{1}{1 - f_1^* f_2^*} \begin{bmatrix} f_2^*(1 - f_1^*) & 1 - f_2^* \\ 1 - f_1^* & f_1^*(1 - f_2^*) \end{bmatrix}, \tag{8}$$

which is an alternating renewal process. Cox and Miller (1965, p. 353) give the specific solution for the negative exponential distribution which is easily found by substituting (7) and (8). For simplicity we can set $\rho_1 = \rho_2 = 1$ and find

$$P_{(t)} = \begin{bmatrix} \Pi_2 - \Pi_2 e^{-t} & \Pi_1 + \Pi_2 e^{-t} \\ \Pi_2 + \Pi_1 e^{-t} & \Pi_1 - \Pi_1 e^{-t} \end{bmatrix} . \qquad (9)$$

Comparing (9) with (6) shows that the transition probabilities created by merging multistory lithologies once again converge towards the stable probability vector, when t becomes large. With t equal to zero, however, transitions from one to one and from two to two have zero probability and consequently $P_{12} = P_{21} = 1.0$.

The last findings have an important bearing on the procedures which one adopts for the analysis of stratigraphic sections. The semi-Markov process can be recovered only if a section is treated as a series of point events, whereby an event is the completion of each bed. A short review of the traditional methods of analysis will show that none of them is applicable to the semi-Markov process.

Vistelius (1949) structured his sections by choosing natural units, i.e. sedimentary beds. Multistory lithologies are fully recognized and, providing that the units are geologically meaningful as is doubtlessly true for deposits laid down by turbidity currents, the method should recover the embedded chain.

Alternatively, sections have been divided into equal thickness increments (Krumbein, 1967) and each increment is treated as a lithological unit. This method provides thickness information but completely loses the events of completing the natural beds. The effect will be similar to ignoring multistory lithologies in the semi-Markov model. As can be seen from (9), the embedded Markov chain enters into this process only by its stable probability vector and this is the only information which can be extracted.

The method of deliberately ignoring multistory lithologies not only loses the thickness information but also the stable probability vector of the embedded matrix; in this context it is difficult to see any geological meaning to this approach. To illustrate the analytical methods, a series of 1000 beds with exponential thickness distribution was generated using (5) as embedded matrix. The series could be regarded as a sandstone-shale series in which the sandstones have an average thickness of 3.0 units and the shale of 1.5 units. Two computer programs have been tested. The first is a standard Markov chain program in which the section is subdivided into equal increments. The size of this step has to be small compared with the average thickness of beds and was taken as 0.1 units. The program calculates first- and higher order transition probabilities. The higher order probabilities test the type of lithology encountered at a distance which is a multiple of the increment and in this manner it is equivalent to the $p(t)$ function. The program cannot differentiate between whether the artificial steps have returned into the same bed or into a different bed with the identical lithology. It therefore treats the section as if multi-story lithologies have merged and we obtain results in accordance with (9) (Fig. 2). Attention has to be given to the definition of the semi-Markov process in which a point event of type i marks the completion of a bed of type i. This program has tested for lithologies (i.e. the type of interval) and not for the completion of the event. The effect is simply that one has to interchange the rows in the $P(t)$ matrix to compare the results with (9) and this has been done in Figure 2.

The experiment illustrates the well-known weakness of the equal increment method. If one chooses a small increment, the leading diagonal of the transition matrix becomes almost 1.0; if one chooses large steps, all detail is lost. The method, and this perhaps has not been sufficiently stressed in the earlier literature, has its main use in investigating the oscillating or cyclical behavior of transition matrices but not for comparing

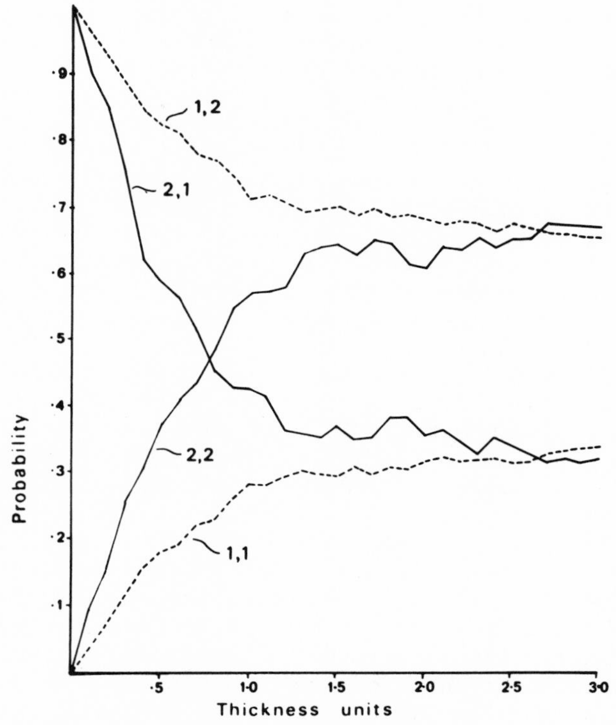

Figure 2. - Transition probabilities at 0.1 unit steps, using Markov-chain program

the probability structure of, say, two formations.

The second experiment was accomplished by modifying the computer program with a single instruction having the effect that a transition was only counted if the intercept under test contained a bedding plane. In this manner, the first program is transformed into a point-event program and, if a bedding plane is encountered, the program searches for the distance at which the next bedding plane of a certain type is found. Obviously the number of transitions to be counted in this manner is smaller than previously and consequently the $p(t)$ function (Fig. 3) is somewhat more irregular. Nevertheless it is clearly seen that this time the result is as predicted by (6).

In the previous section on bed formation, it was concluded that the exponential thickness distribution is only a special situation of the Gamma distribution. Therefore, it is necessary to investigate equation (2) if the Laplace transform of the latter

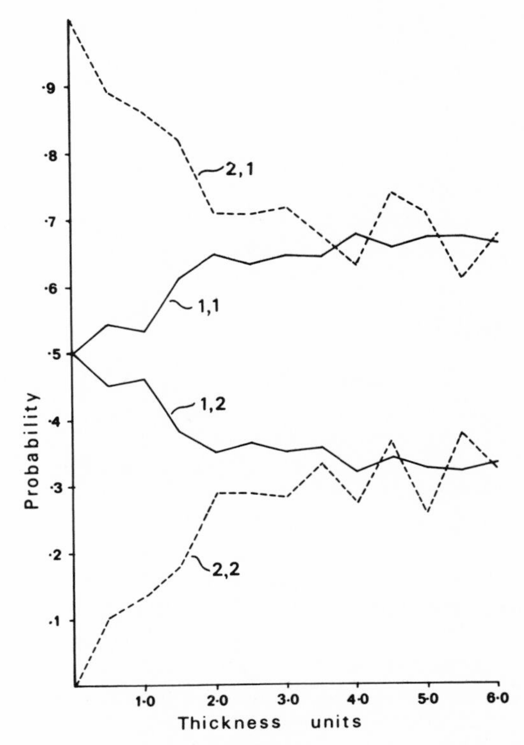

Figure 3. - Transition probabilities at 0.5 unit steps using Point-event program.

distribution is substituted. Calculations involving the Gamma distribution become more and more laborius if the parameter k increases and once again simulation methods seem to be indicated. The simple case of $k = 2$ in the alternating process which is of practical value, may be solved by analytical methods. The transform

$$\frac{\rho_1^2}{(\rho_1 + s)^2} \quad \text{and} \quad \frac{\rho_2^2}{(\rho_2 + s)^2}$$

are therefore substituted into equation (7) and

$$sP_{11}(s) = \frac{\rho_2^2(\rho_1 + s)^2 - \rho_1^2\rho_2^2}{(\rho_1 + s)^2(\rho_2 + s)^2 - \rho_1^2\rho_2^2} \tag{10}$$

is obtained which can be simplified to

$$P_{11}(s) = \frac{\rho_2^2}{s} \frac{2\rho_1 + s}{(1 + s)(a + s + s^2)} \quad \text{where } a = 2\rho_1\rho_2.$$

This can be inverted to

$$P_{11}(t) = A\rho_2^2 \left[(1-e^{-t}) - \frac{e^{-t/2}\sin Bt}{B}\right] + \rho_2^2 \left[\frac{1}{a} - \frac{e^{-t/2}}{\sqrt{a}B}\sin(Bt + \tan^{-1}\sqrt{4a-1})\right] \tag{11}$$

where, $A = \dfrac{\rho_1 - \rho_2}{a}$ and $B = \dfrac{\sqrt{4a - 1}}{2}$ when $4a > 1$.

The transition probabilities are again exponential but with a superposed, strongly damped sine wave. Expression (11) gives only the leading term $p_{11}(t)$ of the transition matrix, the term is zero when $t = 0$, and it approaches ρ_2 when t becomes large. It thus corresponds to the equivalent term in equation (9). The same correspondence applies to the remaining three components.

When the parameter k is low, the damping of the sine wave is pronounced. To illustrate the phenomenon more clearly, a series was generated for $k = 16$, the transition probabilities were calculated by the first program (Fig. 4). This process produces a stratigraphic record which by geological standards would be called cyclical (Schwarzacher, 1969) although the environmental history can be fully described by a nonoscillating Markov chain.

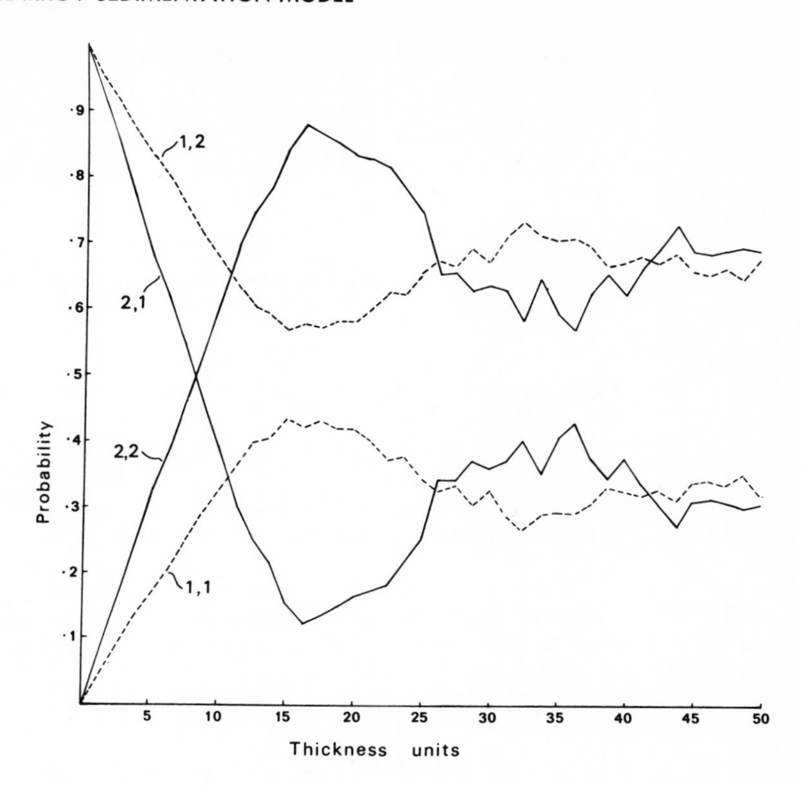

Figure 4. - Transition probabilities of Gamma distributed semi-Markov process.

THICKNESS DISTRIBUTION OF BEDS

The thickness distribution of beds generated by the semi-Markov process should be considered. No problem exists if all multistory lithologies are recorded as the thickness distribution is specified by $f(x)$. However, if multistory lithologies are not recorded, identical lithologies will merge to form a single bed. The probability of a return to an identical state is given by the powers of the elements in the embedded matrix. The final thickness distribution will be the sum of the frequency distributions having a mean which is 1, 2, 3, ... times the mean of $f(x)$ and the total area of which equals to the 1, 2, 3, ... power of a_{ii}. The thickness distribution (Fig. 5) is derived from the previously simulated series with $k = 16$, whereby ρ of the Gamma distribution $f(x)$ was taken as one. Maxima should occur at $(k-1)$, $2(k-1)$,..., i.e. at 15, 30, 45 ... The semi-Markov process therefore will

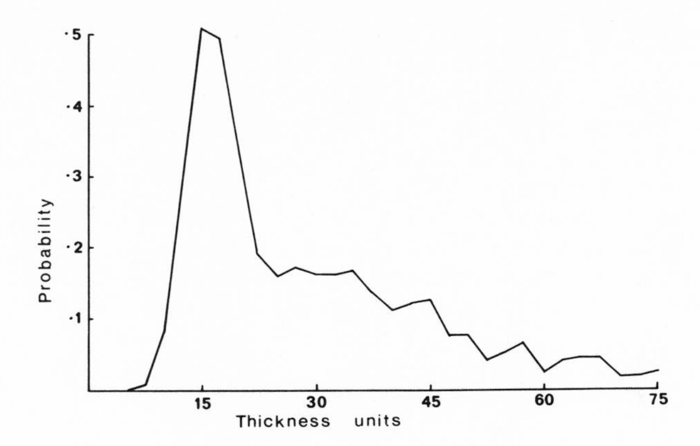

Figure 5. - Thickness distribution of beds generated by Gamma distributed semi-Markov process.

generate polymodal thickness distributions although the subsidiary exception arises if $f(x)$ is an exponential distribution, then the distribrution of merged beds is also exponential.

Let $\mu_{ij} = \frac{1}{\rho_{ij}}$ denote the four means of $f(x)_{ij}$ in the two state system, $f(x)$ being exponential. A single bed of state 1 is formed if a_{21} is immediately succeeded by a_{12} the contributions of 1, 2, ... n layers can be written

$$f^{(1)} = \mu_{21}\, a_{21}\, a_{12}$$

$$f^{(2)} = \mu_{21}\, a_{21}\, a_{12}\, \mu_{11}\, a_{11}$$

$$f_n^{(\mu)} = \mu_{21}\, a_{21}\, a_{12}\, (\mu_{11}a_{11})^{n-1} .$$

The expected length of the process being in state 1 can be written as

$$E_{(T)} = \sum_{n=1}^{\infty} nf^{(n)} = \frac{\mu_{21}a_{21}a_{12}}{(1-\mu_{11}a_{11})^2} . \tag{12}$$

Using (5) as matrix A and setting $\mu_{ij} = 1$ expression (12) simplifies to $\frac{1}{a_{12}}$, and the thickness distribution of the merged bed can be written as

$$f(x)_1 = a_{12}e^{-a_{12}t}$$

(an identical result obtained by Dacey and Krumbein, 1970). In the situation where the parameters of $f(x)$ are known, the embedded Markov chain may be recovered from the thickness distributions of multistory beds and analysis of the bed-thickness distribution becomes an important compliment to the investigation of the transition probabilities. Of course, generally the parameters of bed formation will not be known, but comparative studies of sedimentation may eventually give valuable information.

EXAMPLES OF OBSERVED BED-THICKNESS DISTRIBUTIONS

It is widely accepted that bed thickness has a lognormal distribution. This traditional belief is based on the fact that when data are plotted, they give reasonably straight lines on lognormal probability paper or Chi-square tests indicate a good fit. Any rigorous statistical test against the possibility of an alternative skew distribution would require a large amount of data of a quality which is seldom obtained from geological measurements. The lognormal distribution then has been adopted for convenience and for its usefulness in statistical analysis but as yet, no theory of bed formation exists which leads to this distribution. The proposed Gamma distribution on the other hand, is similar to the lognormal distribution within certain ranges and also provides a good fit to observed data. Compared with the lognormal distribution, the Gamma distribution has a shorter tail and gives therefore a particularly good description of the relatively thicker beds. It has been noted (Nederlof, 1959) that the thicker beds usually depart from the lognormal distribution law.

The following three examples of bed-thickness data were taken from a Carboniferous limestone-shale sequence (Schwarzacher, 1964) and represent three stratigraphic horizons (in ascending order). The thickness frequency distributions of limestone beds and fitted Gamma distributions are given in Figure 6. The Benbulbin Shale is represented by 137 measurements, giving a Gamma distribution which is nearly exponential and with the parameters $k = 1.16$ and $\rho = 0.2381$. The Glencar Limestone ($n = 206$) gave $k = 2.53$ and $\rho = 0.4363$ and the Dartry Limestone ($n = 175$) $k = 6.56$ and $\rho = 0.1978$. The last distribution is almost symmetrical. The parameter k, in each of these examples is based on a maximum likelihood estimate found by an iterative procedure; the method is briefly explained in Cox and Lewis (1966). The examples illustrate the excellent fit which can be obtained by the Gamma distribution. The distribution has not only the advantage that its genesis can be readily understood, but also provides a family of curves which

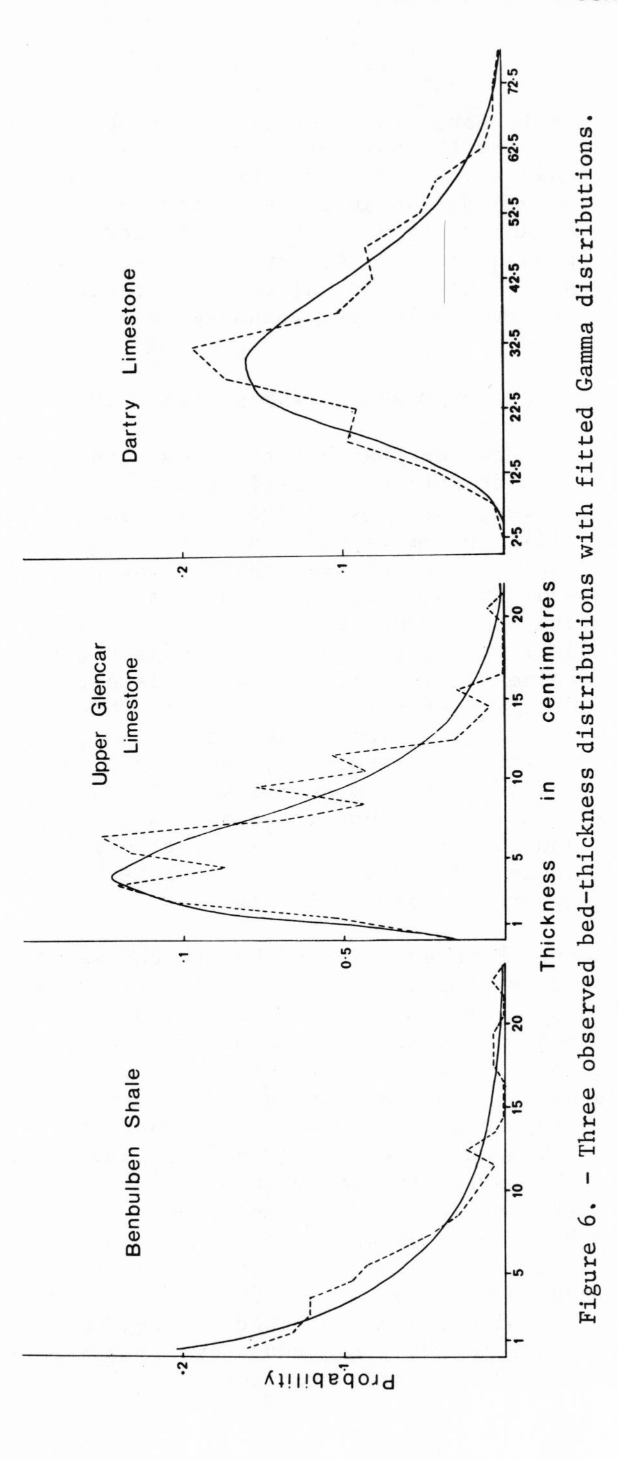

Figure 6. - Three observed bed-thickness distributions with fitted Gamma distributions.

by the change of a single parameter comprises the exponential to the normal distribution. From a sedimentological point of view, it will be extremely interesting to determine whether a physical meaning can be attributed to the parameters of the bed-thickness distributions. The concept of the number of depositional stages per bed seems to explain some known facts. For example, Fiege (1937) plotted the thickness distributions of beds that consisted of different grain sizes. If the grain size is large, very asymmetrical distributions result and the parameter k is consequently low. Beds consisting of a clay-type sediment resulted in more symmetrical distributions and beds might be accepted as consisting of a greater number of individual steps. It has been argued previously that it is unlikely that one-step beds will be found. This explains the relative rarity of the exponential thickness distributions for which Krumbein and Dacey (1969) have been searching, but which seem to be nevertheless the fundamental thickness distribution.

The polymodal thickness distributions, predicted from the semi-Markov model, may occur more frequently than realized at present. However, the damping, which is caused by the embedded chain, may be difficult to recognize in subsidiary peaks and precise measurements are needed to show this phenomenon. Measure-

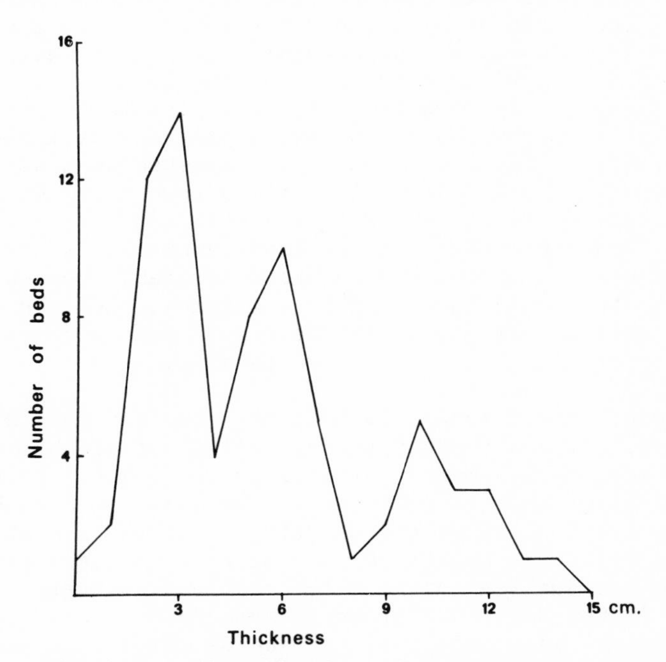

Figure 7. - Thickness distribution of upper Glencar Limestone.

ments were taken from a 10-meter peel of sections of the Glencar Limestone. A bed was defined as any thickness of limestone which was not interrupted by a bedding plane when the record was scrutinized under the microscope. Although the sequence contains only 62 measurements (Fig. 7), the strict definition of beds together with the accuracy of measurements resolve the thickness distribution into three modes which are repeated at 3 centimeter intervals. The exponential decrease of the peaks also can be seen clearly. It is hoped to give a full analysis of this example later.

CONCLUDING REMARKS

The analysis of stratigraphic sections be means of Markov models is undergoing continuous development (Krumbein and Dacey, 1969). The basic model proposed a series of (environmental) events which lead to discrete units of sedimentation. Both the events themselves and the resulting sequence of rock strata, were assumed to be Markov chains. Difficulties lie in structuring the sequence of rock strata. A.B. Vistelius (lecture in London in 1970) demonstrated that the concept of a simple Markov chain can be maintained if the correct group of lithologies are combined to form a natural stratigraphic step. If the stratigraphic record is considered the result of environmental developments and sedimentation processes, the grouping of lithologies must be caused by what has been called in this paper, the mechanism of bed formation. Of course, the fact that a first-order Markov component can be split from this record does not prove in itself that the environmental history is a simple Markov chain. However, progress will have been made if the stratigraphic record can be interpreted as a simple Markov chain together with a mechanism of bed formation which makes sense in sedimentological terms. On the other hand, if geological considerations indicate a more complex environmental history (for example, cyclic sedimentation) and nature is probably not sufficiently obliging to remain Markovian at all times, then the model may have to be more complicated. It seems good policy to retain the simpler concepts as long as they are tenable.

The choice of Markov models is becoming more and more a question of sedimentology and related fundamental stratigraphic thinking. Thus the semi-Markov model arises from the quantified time scale of the stratigrapher in contrast to the continuous time scale which may provide the better description of sedimentation processes. The following questions may occur to the sedimentologist. What are the smallest steps which can be relatively dated in any sediment? Why does bed formation seem to be a discrete process and what are the geological thresholds which make sedimentation and possibly environmental developments move in steps?

Selley (1970) writes about probabilistic models: "Interesting and impressive as these methods are, they are of relatively little value to the geologist seeking to understand the vertical arrangements of facies in a sequence in an attempt to diagnose the environments". There may indeed be some truth in this statement. If these "interesting and impressive methods" actually help geologists to realize what is involved in "seeking to understand", then they may prove to be more than of "relatively little value".

ACKNOWLEDGMENTS

I am grateful to Dr. A. P. Gallagher for his help with the inversion of Laplace transforms. Prof. A. Williams very kindly read the manuscript.

REFERENCES

Carr, D. D., and others, 1966, Stratigraphic sections, bedding sequences, and random processes: Science, v. 154, no. 3753, p. 1162-1164.

Cox, D. R., and Lewis, P. A. W., 1966, The statistical analysis of series of events: Methuen and Co., London, 285 p.

Cox, D. R., and Miller, H. D., 1965, The theory of stochastic processes: John Wiley & Sons, New York, 398 p.

Dacey, M. F., and Krumbein, W. C., 1970, Markovian models in stratigraphic analysis: Jour. Intern. Assoc. Math. Geology, v. 2, no. 2, p. 175-191.

Fiege, K., 1937, Untersuchungen uber zyklische Sedimentation geosynklinaler und epikontinentaler Raume: Abh. preussische Geol. Landesanst., N. F. Heft 177, p. 1-218.

Jopling, A. V., 1964, Interpreting the concept of the sedimentation unit: Jour. Sed. Pet., v. 34, no.1 , p. 165-172.

Kolmogorov, A. N., 1951, Solution of a problem in probability theory connected with the problem of the mechanism of stratification: Am. Math. Soc. Trans., no. 53, 8 p.

Krumbein, W. C., 1967, FORTRAN IV computer programs for Markov chain experiments in geology: Kansas Geol. Survey Computer Contr. 13, 38 p.

Krumbein, W. C., and Dacey, M. F., 1969, Markov chains and embedded Markov chains in geology: Jour. Intern. Assoc. Math. Geology, v. 1, no. 1, p. 79-94.

Nederlof, M. H., 1959, Structure and sedimentology of the Upper Carbonfierous of the upper Pisuerga Valleys, Cantabrian Mountains, Spain: Leidse Geol. Med., v. 24, p. 603-703.

Otto, G. H., 1938, The sedimentation unit and its use in field sampling: Jour. Geology, v. 46, p. 569-582.

Schwarzacher, W., 1964, An application of statistical time-series analysis of a limestone-shale sequence: Jour. Geology, v. 72, no. 2, p. 195-213.

Schwarzacher, W., 1968, Experiments with variable sedimentation rates, in Computer applications in the earth sciences: Colloquium on simulation, Kansas Geol. Survey Computer Contr. 22, p. 19-21.

Schwarzacher, W., 1969, The use of Markov chains in the study of sedimentary cycles: Jour. Intern. Assoc. Math. Geology, v. 1, no. 1, p. 17-39.

Selley, R. C., 1970, Studies of sequence in sediments using a simple mathematical device: Jour. Geol. Soc. London, v. 125, p. 557-581.

Vistelius, A. B., 1949, On the question of the mechanism of the formation of strata: Doklady Akad. Nauk SSSR, v. 65, p. 191-194.

Vistelius, A. B., and Feigel'son, T. S., 1965, On the theory of bed formation: Doklady Akad. Nauk SSSR, v. 164, p. 158-160.

INDEX

algal-bank growth, 16
alluvial channels, 23
American Midcontinent, 28
Appalachian Basin, 223
aquifers, 55
Aquitaine Basin, 53, 61
artificial flow, 67
Atterberg cylinder, 191
autoassociation, 31, 36, 40, 41, 49
autocorrelation methods, 31
bed formation, 251
bed-thickness distribution, 247
Benbulbin Shale, 263
bivariate correlation, 239
boundary conditions, 58, 193
Britain, 205, 209, 225
British Coal Measures, 223
canonical correlation, 235, 239
Carboniferous sections, 205, 225, 263
Chemery structure, 140, 142
Cherokee Group, 224
Chi-square tests, 263
classification, 90, 97, 100
closure of open systems, 169
cluster analysis, 15, 33, 82, 205, 210, 212
  single linkage, 210, 225
  weighted pair-group methods, 91, 100
  unweighted pair-group method, 100
  unweighted average linkage, 210, 221, 225
coarseness index, 83
coefficients
  linearity, 87
  saturation, 72
  similarity, 225
  variation, 110
conditional laws, 149
conditional probabilities, 31
conditional simulation, 147
continuity equation, 55
correlation coefficients, 86, 100, 169, 210, 221, 234
covariance partitioning, 170
creeping flow, 199
Cretaceous-Paleogene flysch, 117
critical-diagram fields, 126
crossassociation, 204
cyclothems, 28, 38, 48, 139, 203, 266
  typical, 28
  idealized, 28
Damacusa Valley, 118
Darcy's Law, 54
Dartry Limestone, 263
data analysis, 12
decision variables, 15
deltaic cycles, 204, 221
deltas, 22, 233
dendrograms, 92, 107, 210, 221
depositional environment, 82, 221
diagenetic processes, 191
diffusion model, 125
discrete Markov chain, 249
discriminant-function analysis, 15
dissolving particle, 194
distance coefficient, 89, 100
East Carpathians, 115, 118
eigenvalues, 86
eigenvectors, 86, 240
embedded Markov chain, 249
estuarine sedimentation, 16
Euclidean distance, 87, 89
evaporite sedimentation, 16
exponential distribution, 251
exponential law, 133
facies
  attributes, 108
  models, 81, 98
factor analysis, 15, 235
Fick's first law, 193
flow rates, 58
flysch deposits, 115
Fourier series, 13, 31
gamma distribution, 253

geometry of rivers, 3
Glencar Limestone, 263
grain-size
  moment, 126
  reduction, 192, 196
hangover limit, 3
hierarchical grouping, 96
Honaker Trail Formation, 83
hydraulic parameters, 20
hydrologic processes, 53
inherent correlation, 170
inverse method, 65, 66, 71
Kansas, 30, 39
Kincardine Basin, 205, 209, 210, 221
kinematic viscosity, 195
Laplace transform, 253, 259
laws of statics, 2
lead and zinc, 233, 234
least squares, 13, 168
Le Chatelier's principle, 2
Levallois' clays, 141
Limestone Coal Group, 206, 221
limit partition, 100
lithology
  correlation, 152
  states, 30, 33
  successions, 27
lognormal distribution law, 263
Mansfield Marine Band, 224
marine regression, 28
marine transgression, 28
Markov property, 16, 31, 36, 204
mathematical search procedures, 81
Medoc vineyards, 62
megacyclothems, 28, 38, 48
minimization principle, 2
Mississippi River, 131
models
  deterministic, 16
  digital, 54
  process, 122
  random, 140
  regression, 140
  response, 122
  simple partitioning, 169
  simulation, 16
  transgression, 140
Monte Carlo simulation, 115, 117, 171
multiple regression, 236
multistory lithologies, 205, 250, 257
nearest-neighbor algorithm, 100, 104
Niger Delta, 236
objective criterion, 9
objective function, 15
Oligocene marls, 70
optimization criteria, 1
Paris Basin, 141
Pennsylvania, 225
Pennsylvanian System, 30
phenon line, 96
piezometric heads, 63, 67
Pliocene aquifer, 70
Polomet River, 135
power law, 13
Prandtl's theory, 195
prediction, 16
principal components, 86, 90, 235, 243
principle of least restraint, 2
probability laws, 144
processes of dissolution, 191
random walk, 251
recurved spit, 16
Rhine alluvial aquifer, 70
sampling procedures, 5
sand
  dunes, 175
  movement, 176
  surface, 180
San Juan River, 83

Sbrancani Valley, 118
Scotland, 206, 223
sediment deposition, 132
semi-Markov process, 247, 248, 253
set-functions, 66
similarity measures, 88
solution rates, 191
standardization, 85
statistical correlation, 167
stochastic components, 16, 116
stochastic variables, 116
storage coefficients, 57, 66
subfacies, 111
substitutability analysis, 27, 34, 41, 49
suspended-sediment transport, 17
suspended-sediment discharge, 20, 23
taxonomic distance, 225
Tertiary aquifers, 61
thickness distribution of beds, 261
time-series analysis, 31
transition probabilities, 31, 34, 205, 209, 225, 257
transmissivity, 56, 66
transport of suspended sediments, 128
trend-surface coefficients, 169
trial and error methods, 61
turbulent flow, 125
two-state system, 48, 254
Upper Carboniferous strata, 28
Upper Paleozoic succession, 30
United States, 205, 209, 224, 225
velocity profile, 18
Vinetisu Valley, 118
water flow in aquifers, 54